Text Mining in Practice with R

Text Mining in Practice with R

Ted Kwartler

Registered Office
John Wiley & Sons Ltd, The Atrium, Southern Gate, Chichester, West Sussex, PO19 8SQ, UK

Editorial Offices
111 River Street, Hoboken, NJ 07030, USA
9600 Garsington Road, Oxford, OX4 2DQ, UK
The Atrium, Southern Gate, Chichester, West Sussex, PO19 8SQ, UK

For details of our global editorial offices, customer services, and more information about Wiley products visit us at www.wiley.com.

Wiley also publishes its books in a variety of electronic formats and by print-on-demand. Some content that appears in standard print versions of this book may not be available in other formats.

Library of Congress Cataloging-in-Publication Data
Names: Kwartler, Ted, 1978- author.
Title: Text mining in practice with R / Ted Kwartler.
Description: Hoboken, NJ : John Wiley & Sons, 2017. | Includes
   bibliographical references and index.
Identifiers: LCCN 2017006983 (print) | LCCN 2017010584 (ebook) | ISBN
   9781119282013 (cloth) | ISBN 9781119282099 (pdf) | ISBN 9781119282082
   (epub)
Subjects: LCSH: Data mining. | Text processing (Computer science)
Classification: LCC QA76.9.D343 K94 2017 (print) | LCC QA76.9.D343 (ebook) |
   DDC 006.3/12--dc23
LC record available at https://lccn.loc.gov/2017006983

Cover Design: Wiley
Cover Image: © ChrisPole/Gettyimages

Set in 10/12pt WarnockPro by SPi Global, Chennai, India

10 9 8 7 6 5 4 3 2 1

"It's the math of talking…your two favorite things!"

This book is dedicated to my beautiful wife, and best friend Meghan. Your patience, support and assurance cannot be quantified.

Additionally Nora and Brenna are my motivation, teaching me to be a better person.

Contents

Foreword

This book has been a long labor of love. When I agreed to write a book, I had no idea of the amount of work and research needed. Looking back, it was pure hubris on my part to accept a writing contract from the great people at Wiley. The six-month project extended outward to more than a year! From the outset I decided to write a book that was less technical or academic and instead focused on code explanations and case studies. I wanted to distill my years of work experience, blog reading and textbook research into a succinct and more approachable format. It is easy to copy a blog's code or state a textbook's explanation verbatim, but it is wholesale more difficult to be original, to explain technical attributes in an easy-to-understand manner and hopefully to make the journey more fun for the reader.

Each chapter demonstrates a text mining method in the context of a real case study. Generally, mathematical explanations are brief and set apart from the code snippets and visualizations. While it is still important to understand the underlying mathematical attributes of a method, this book merely gives you a glimpse. I believe it is easier to become an impassioned text miner if you get to explore and create first. Applying algorithms to interesting data should embolden you to undertake and learn more. Many of the topics covered could be expanded into a standalone book, but here they are related as a single section or chapter. This is on purpose, so you get a quick but effective glimpse at the text mining universe! So my hope is that this book will serve as a foundation as you continually add to your data science skillset.

As a writer or instructor I have always leaned on common sense and non-academic explanations. The reason for this is simple: I do not have a computer science or math degree. Instead, my MBA gives me a unique perspective on data science. It has been my observation that data scientists often enjoy the modeling and data wrangling, but very often fail to completely understand the needs of the business. Thus many data science business applications are actually months in implementation or miss a crucial aspect. This book strives to have original and thought-provoking case studies with truly messy data. In other text mining or data science books, data that perfectly describes the method is illustrated so the concept can be understood. In this book, I reverse that approach and attempt to use real data in context so you can learn how typical text mining data is modeled and what to expect. The results are less pretty but more indicative of what you should expect as a text mining practitioner.

"It takes a village to write a book."

Throughout this journey I have had the help of many people. Thankfully, family and friends have been accommodating and understanding when I chose writing ahead of social gatherings. First and foremost thanks to my mother, Trish, who gave me the gift of gab, and qualitative understanding and to my father Yitz, who gave me quantitative and technical writing acumen. Additional thanks to Paul, MaryAnn, Holly, Rob, K, and Maureen for understanding when I had to steal away and write during visits.

Thank you to Barry Keating, Sarv Devaraj and Timothy Gilbride. The Notre Dame family, with their supportive, entertaining professors put me onto this path. Their guidance, dialogue and instructions opened my eyes to machine learning, data science and ultimately text mining. My time at Notre Dame has positively affected my life and those around me. I am forever grateful.

Multiple data scientists have helped me along the way. In fact to many to actually list. Particular thanks to Greg, Zach, Hamel, Jeremy, Tom, Dalin, Sergey, Owen, Peter, Dan, Hugo and Nick for their explanations at different points in my personal data science journey.

This book would not have been possible if it weren't for Kathy Powers. She has been a lifelong friend and supporter and amazingly stepped up to make revisions when asked. When I changed publishers and thought of giving up on the book her support and patience with my poor grammar helped me continue. My entire family owes you a debt of gratitude that is never able to be repaid.

1

What is Text Mining?

In this chapter, you will learn

- the basic definition of practical text mining
- why text mining is important to the modern enterprise
- examples of text mining used in enterprise
- the challenges facing text mining
- an example workflow for processing natural language in analytical contexts
- a simple text mining example
- when text mining is appropriate

Learning how to perform text mining should be an interesting and exciting journey throughout this book. A fun artifact of learning text mining is that you can use the methods in this book on your own social media or online exchanges. Beyond these everyday online applications to your personal interactions, this book provides business use cases in an effort to show how text mining can improve products, customer service, marketing or human resources.

1.1   What is it?

There are many technical definitions of text mining both on the Internet and in textbooks, but as the primary goal of text mining in this book is the extraction of an output that is useful such as a visualization or structured table of outputs to be used elsewhere; this is my definition:

> Text mining is the process of distilling actionable insights from text.

Text mining within the context of this book is a commitment to real world cases which impact business. Therefore, the definition and this book are aimed

Text Mining in Practice with R, First Edition. Ted Kwartler.
© 2017 John Wiley & Sons Ltd. Published 2017 by John Wiley & Sons Ltd.

at meaningful distillation of text with the end goal to aid a decision-maker. While there may be some differences, the terms text mining and text analytics can be used interchangeably. Word choice is important; I use text mining because it more adequately describes the uncovering of insights and the use of specific algorithms beyond basic statistical analysis.

1.1.1 What is Text Mining in Practice?

In this book, text mining is more than an academic exercise. I hope to show that text mining has enterprise value and can contribute to various business units. Specifically, text mining can be used to identify actionable social media posts for a customer service organization. It can be used in human resources for various purposes such as understanding candidate perceptions of the organization or to match job descriptions with resumes. Text mining has marketing implications to measure campaign salience. It can even be used to identify brand evangelists and impact customer propensity modeling. Presently the state of text mining is somewhere between novelty and providing real actionable business intelligence. The book gives you not only the tools to perform text mining but also the case studies to help identify practical business applications to get your creative text mining efforts started.

1.1.2 Where Does Text Mining Fit?

Text mining fits within many disciplines. These include private and academic uses. For academics, text mining may aid in the analytical understanding of qualitatively collected transcripts or the study of language and sociology. For the private enterprise, text mining skills are often contained in a data science team. This is because text mining may yield interesting and important inputs for predictive modeling, and also because the text mining skillset has been highly technical. However, text mining can be applied beyond a data science modeling workflow. Business intelligence could benefit from the skill set by quickly reviewing internal documents such as customer satisfaction surveys. Competitive intelligence and marketers can review external text to provide insightful recommendations to the organization. As businesses are saving more textual data, they will need to break text-mining skills outside of a data science team. In the end, text mining could be used in any data driven decision where text naturally fits as an input.

1.2 Why We Care About Text Mining

We should care about textual information for a variety of reasons.

- Social media continues to evolve and affect an organization's public efforts.

- Online content from an organization, its competitors and outside sources, such as blogs, continues to grow.
- The digitization of formerly paper records is occurring in many legacy industries, such as healthcare.
- New technologies like automatic audio transcription are helping to capture customer touchpoints.
- As textual sources grow in quantity, complexity and number of sources, the concurrent advance in processing power and storage has translated to vast amounts of text being stored throughout an enterprise's data lake.

Yet today's successful technology companies largely rely on numeric and categorical inputs for information gains, machine learning algorithms or operational optimization. It is illogical for an organization to study only structured information yet still devote precious resources to recording unstructured natural language. Text represents an untapped input that can further increase competitive advantage. Lastly, enterprises are transitioning from an industrial age to an information age; one could argue that the most successful companies are transitioning again to a customer-centric age. These companies realize that taking a long term view of customer wellbeing ensures long term success and helps the company to remain salient. Large companies can no longer merely create a product and forcibly market it to end-users. In an age of increasing customer expectations customers want to be heard by corporations. As a result, to be truly customer centric in a hyper competitive environment, an organization should be listening to their constituents whenever possible. Yet the amount of textual information from these interactions can be immense, so text mining offers a way to extract insights quickly.

Text mining will make an analyst's or data scientist's efforts to understand vast amounts of text easier and help ensure credibility from internal decision-makers. The alternative to text mining may mean ignoring text sources or merely sampling and manually reviewing text.

1.2.1 What Are the Consequences of Ignoring Text?

There are numerous consequences of ignoring text.

- Ignoring text is not an adequate response of an analytical endeavor. Rigorous scientific and analytical exploration requires investigating sources of information that can explain phenomena.
- Not performing text mining may lead an analysis to a false outcome.
- Some problems are almost entirely text-based, so not using these methods would mean significant reduction in effectiveness or even not being able to perform the analysis.

Explicitly ignoring text may be a conscious analyst decision, but doing so ignores text's insightful possibilities. This is analogous to an ostrich that sticks its head in the ground when confronted. If the aim is robust investigative quantitative analysis, then ignoring text is inappropriate. Of course, there are constraints to data science or business analysis, such as strict budgets or time-lines. Therefore, it is not always appropriate to use text for analytics, but if the problem being investigated has a text component, and resource constraints do not forbid it, then ignoring text is not suitable.

Wisdom of Crowds

As an alternative, some organizations will sample text and manually review it. This may mean having a single assessor or panel of readers or even outsourcing analytical efforts to human-based services like `mturk` or `crowdflower`. Often communication theory does not support these methods as a sound way to score text, or to extract meaning. Setting aside sampling biases and logistical tabulation difficulties, communication theory states that the meaning of a message relies on the recipient. Therefore a single evaluator introduces biases in meaning or numerical scoring, e.g. sentiment as a numbered scale. Additionally, the idea behind a group of people scoring text relies on Sir Francis Galton's theory of "Vox Populi" or wisdom of crowds.

To exploit the wisdom of crowds four elements must be considered:
- Assessors need to exercise independent judgments.
- Assessors need to possess a diverse information understanding.
- Assessors need to rely on local knowledge.
- There has to be a way to tabulate the assessors' results.

Sir Francis Galton's experiment exploring the wisdom of crowds met these conditions with 800 participants. At an English country fair, people were asked to guess the weight of a single ox. Participants guessed separately from each other without sharing the guess. Participants were free to look at the cow themselves yet not receive expert consultation. In this case, contestants had a diverse background. For example, there were no prerequisites stating that they needed to be a certain age, demographic or profession. Lastly, guesses were recorded on paper for tabulation by Sir Francis to study. In the end, the experiment showed the merit of the wisdom of crowds. There was not an individual correct guess. However, the median average of the group was exactly right. It was even better than the individual farming experts who guessed the weight.

If these conditions are not met explicitly, then the results of the panel are suspect. This may seem easy to do, but in practice it is hard to ensure within an organization. For example a former colleague at a major technology company in California shared a story about the company's effort to create Internet-connected eyeglasses. The eyeglasses were shared with internal employees, and feedback was then solicited. The text feedback was sampled and scored by internal employees. At first blush this seems like a fair assessment of the product's features and expected popularity. However, the conditions for the wisdom of crowds were not met. Most notably, the need for a decentralized understanding of the question was not met. As members of the same technology company, the respondents are already part of a self-selected group that understood the importance of the overall project within the company. Additionally, the panel had a similar assessment bias because they were from the same division that was working on the project. This assessing group did not satisfy the need for independent opinions when assessing the resulting surveys. Further, if a panel is creating summary text as the output of the reviews, then the effort is merely an information reduction effort similar to numerically taking an average. Thus it may not solve the problem of too much text in a reliable manner. Text mining solves all these problems. It will use all of the presented text and does so in a logical, repeatable and auditable way. There may be analyst or data scientist biases but they are documented in the effort and are therefore reviewable. In contrast, crowd-based reviewer assessments are usually not reviewable.

Despite the pitfalls of ignoring text or using a non-scientific sampling method, text mining offers benefits. Text mining technologies are evolving to meet the demands of the organization and provide benefits leading to data-driven decisions. Throughout this book, I will focus on benefits and applied applications of text mining in business.

1.2.2 What Are the Benefits of Text Mining?

There are many benefits of text mining including:

- Trust is engendered among stakeholders because little to no sampling is needed to extract information.
- The methodologies can be applied quickly.
- Using R allows for auditable and repeatable methods.
- Text mining identifies novel insights or reinforces existing perceptions based on all relevant information.

Interestingly, text mining first appears in the Gartner Hype Cycle in 2012. At that moment, it was listed in the "trough of disillusionment." In subsequent years, it has not been listed on the cycle at all, leading me to believe that text analysis is either at a steady enterprise use state or has been abandoned by enterprises as not useful. Despite not being listed, text mining is used across industries and in various manners. It may not have exceeded the over-hyped potential of 2012's Gartner Hype Cycle, but text is showing merit. Hospitals use text mining of doctors' notes to understand readmission characteristics of patients. Financial and insurance companies use text to identify compliance risks. Retailers use customer service notes to make operational changes when

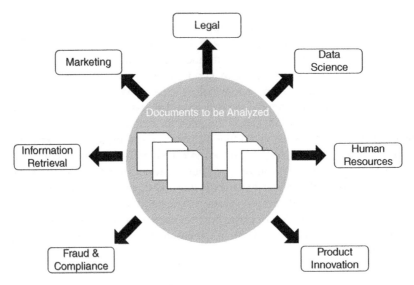

Figure 1.1 Possible enterprise uses of text min.

failing customer expectations. Technology product companies use text mining to seek out feature requests in online reviews. Marketing is a natural fit for text analysis. For example, marketing companies monitor social media to identify brand evangelists. Human resource analytics efforts focus on resume text to match to job description text. As described here, mastering text mining is a skill set sought out across verticals and is therefore a worthwhile professional endeavor. Figure 1.1 shows possible business units that can benefit from text mining in some form.

1.2.3 Setting Expectations: When Text Mining Should (and Should Not) Be Used

Since text is often a large part of a company's database, it is believed that text mining will lead to ground-breaking discoveries or significant optimization. As a result, senior leaders in an organization will devote resources to text mining, expecting to yield extensive results. Often specialists are hired, and resources are explicitly devoted to text mining. Outside of text mining software, in this case R, it is best to use text mining only in cases where it naturally fits the business objective and problem definition. For example, at a previous employer, I wondered how prospective employees viewed our organization compared to peer organizations. Since these candidates were outside the organization, capturing numerical or personal information such as age or company-related perspective scoring was difficult. However, there are forums and interview reviews anonymously shared online. These are shared as text so naturally text mining was an appropriate tool. When using text mining, you should prioritize defining the problem and reviewing applicable data, not using an exotic text mining method. Text mining is not an end in itself and should be regarded as another tool in an analyst's or data scientist's toolkit.

Text mining cannot distill large amounts of text to gain an absolute view of the truth. Text mining is part art and part science. An analyst can mislead stakeholders by removing certain words or using only specific methods. Thus, it is important to be up front about the limitations of text mining. It does not reveal an absolute truth contained within the text. Just as an average reduces information for consumption of a large set of numbers, text mining will reduce information. Sometimes it confirms previously held beliefs and sometimes it provides novel insights. Similar to numeric dimension reduction techniques, text mining abridges outliers, low frequency phrases and important information. It is important to understand that language is more colorful and diverse in understanding than numerical or strict categorical data. This poses a significant problem for text miners. Stakeholders need to be wary of any text miner who knows a truth solely based on the algorithms in this book. Rather, the methods in this book can help with the narrative of the data and the problem at hand, or the outputs can even be used in supervised learning alongside numeric data to improve the predictive outcomes. If doing predictive modeling using text, a best practice when modeling alongside non-text data features is to model with and without the text in the attribute set. Text is so diverse that it may even add noise to predictive efforts. Table 1.1 refers to actual use cases where text mining may be appropriate.

Table 1.1 Example use cases and recommendations to use or not use text mining.

Example use case	Recommendation
Survey texts	Explore topics using various methods to gain a respondent's perspective.
Reviewing a small number of documents	Don't perform text mining on an extremely small corpus, as the results and conclusion can be skewed.
Human resource documents	Tread carefully; text mining may yield insights, but the data and legal barriers may make the analysis inappropriate.
Social media	Use text mining to collect (when allowed) from online sources and then apply preprocessing steps to extract information.
Data science predictive modeling	Text mining can yield structured inputs that could be useful in machine learning efforts.
Product/service reviews	Use text mining if the number of reviews is large.
Legal proceeding	Use text mining to identify individuals and specific information.

Another suggestion for effective text mining is to avoid over using a word cloud. Analysts armed with the knowledge of this book should not create a word cloud without a need for it. This is because word clouds are often used without need, and as a result they can actually diminish their impact. However, word clouds are popular and can be powerful in showing term frequency, among other things, such as the one in Figure 1.2, which runs over the text of this chapter. Throwing caution to the wind, it demonstrates a word cloud of terms in Chapter 1. It is not very insightful because, as expected, the terms text and mining are the most frequent and largest words in the cloud!

In fact, word clouds are so popular that an entire chapter is devoted to various types of word clouds that can be insightful. However, many people consider word clouds a cliché, so their impact is fading. Also, word clouds represent

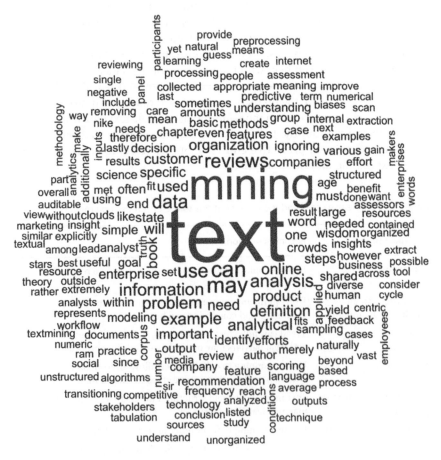

Figure 1.2 A gratuitous word cloud for Chapter 1.

a relatively easy way to mislead consumers of an analysis. In the end, they should be used in conjunction with other methods to confirm the correctness of a conclusion.

1.3 A Basic Workflow – How the Process Works

Text represents unstructured data that must be preprocessed into a structured manner. Features need to be defined and then extracted from the larger body of organized text known as a corpus. These extracted features are then analyzed. The chevron arrows in Figure 1.3 represent structured predefined steps

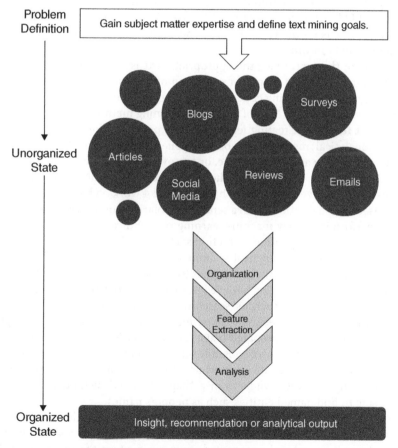

Figure 1.3 Text mining is the transition from an unstructured state to a structured understandable state.

that are applied to the unorganized text to reach the final output or conclusion. Overall Figure 1.3 is a high level workflow of a text mining project.

The steps for text mining include:

1) **Define the problem and specific goals.** As with other analytical endeavors, it is not prudent to start searching for answers. This will disappoint decision-makers and could lead to incorrect outputs. As the practitioner, you need to acquire subject matter expertise sufficient to define the problem and the outcome in an appropriate manner.

2) **Identify the text that needs to be collected.** Text can be from within the organization or outside. Word choice varies between mediums like Twitter and print so care must be taken to explicitly select text that is appropriate to the problem definition. Chapter 9 covers places to get text beyond reading in files. The sources covered include basic web scraping, APIs and R's specific API libraries, like "twitteR." Sources are covered later in the book so you can focus on the tools to text mine, without the additional burden of finding text to work on.

3) **Organize the text.** Once the appropriate text is identified, it is collected and organized into a corpus or collection of documents. Chapter 2 covers two types of text mining conceptually, and then demonstrates some preparation steps used in a "bag of words" text mining method.

4) **Extract features.** Creating features means preprocessing text for the specific analytical methodology being applied in the next step. Examples include making all text lowercase, or removing punctuation. The analytical technique in the next step and the problem definition dictate how the features are organized and used. Chapters 3 and 4 work on basic extraction to be used in visualizations or in a sentiment polarity score. These chapters are not performing heavy machine learning or technical analysis, but instead rely on simple information extraction such as word frequency.

5) **Analyze.** Apply the analytical technique to the prepared text. The goal of applying an analytical methodology is to gain an insight or a recommendation or to confirm existing knowledge about the problem. The analysis can be relatively simple, such as searching for a keyword, or it may be an extremely complex algorithm. Subsequent chapters require more in-depth analysis based on the prepared texts. A chapter is devoted to unsupervised machine learning to analyze possible topics. Another illustrates how to perform a supervised classification while another performs predictive modeling. Lastly you will switch from a "bag of words" method to syntactic parsing to find named entities such as people's names.

6) **Reach an insight or recommendation.** The end result of the analysis is to apply the output to the problem definition or expected goal. Sometimes this can be quite novel and unexpected, or it can confirm the previously held idea. If the output does not align to the defined problem or completely

satisfy the intended goal, then the process becomes repetitious and can be changed at various steps. By focusing on real case studies that I have encountered, I hope to instill a sense of practical purpose to text mining. To that end, the case studies, the use of non-academic texts and the exercises of this book are meant to lead you to an insight or narrative about the issue being investigated. As you use the tools of this book on your own, my hope is that you will remember to lead your audience to a conclusion.

The distinct steps are often specific to the particular problem definition or analytical technique being applied. For example, if one is analyzing tweets, then removing retweets may be useful but it may not be needed in other text mining exploration. Using R for text mining means the processing steps are repeatable and auditable. An analyst can customize the preprocessing steps outlined throughout the book to improve the final output. The end result is an insight, a recommendation or may be used in another analysis. The R scripts in this book follow this transition from an unorganized state to an organized state, so it is important to recall this mental map.

The rest of the book follows this workflow and adds more context and examples along the way. For example, Chapter 2 examines the two main approaches to text mining and how to organize a collection of documents into a clean corpus. From there you start to extract features of the text that are relevant to the defined problem. Subsequent chapters add visualizations, such as word clouds, so that a data scientist can tell the analytical narrative in a compelling way to stakeholders. As you progress through the book the types and methods of extracted features or information grow in complexity because the defined problems get more complex. You quickly divert to covering sentiment polarity so you can understand Airbnb reviews. Using this information you will build compelling visualizations and know what qualities are part of a good Airbnb review. Then in Chapter 5 you learn topic modeling using machine learning. Topic modeling provides a means to understand the smaller topics associated within a collection of documents without reading the documents themselves. It can be useful for tagging documents relating to a subject. The next subject, document classification, is used often. You may be familiar with document classification because it is used in email inboxes to identify spam versus legitimate emails. In this book's example you are searching for "clickbait" from online headlines. Later you examine text as it relates to patient records to model how a hospital identifies diabetic readmission. Using this method, some hospitals use text to improve patient outcomes. In the same chapter you even examine movie reviews to predict box office success. In a subsequent chapter you switch from the basic bag of words methodology to syntactic parsing using the OpenNLP library. You will identify named entities, such as people, organizations and locations within Hillary Clinton's emails. This can be useful in legal proceedings in which the volume of documentation is large and the deadlines are tight. Marketers also use named entity recognition to understand what influencers are discussing. The remaining chapters refocus your attention back to some more basic principles at the top of the workflow, namely where to get text and how to read it into R. This will let you use the scripts in this book with text that is thought provoking to your own interests.

1.4   What Tools Do I Need to Get Started with This?

To get started in text mining you need a few tools. You should have access to a laptop or workstation with at least 4GB of RAM. All of the examples in this book have been tested on a Microsoft's Windows operating systems. RAM is important because R's processing is done "in memory." This means that the objects being analyzed must be contained in the RAM memory. Also, having a high speed internet connection will aid in downloading the scripts, R library packages and example text data and for gathering text from various webpages. Lastly, the computer needs to have an installation of R and R Studio. The operating system of the computer should not matter because R has an installation for Microsoft, Linux and Mac.

1.5   A Simple Example

Online customer reviews can be beneficial to understanding customer perspectives about a product or service. Further, reviewers can sometimes leave feedback anonymously, allowing authors to be candid and direct. While this may lead to accurate portrayals of a product it may lead to "keyboard courage" or extremely biased opinions. I consider it a form of selection bias, meaning that the people that leave feedback may have strong convictions not indicative of the overall product or service's public perception. Text mining allows an enterprise to benchmark their product reviews and develop a more accurate understanding of some public perceptions. Approaches like topic modeling and polarity (positive and negative scoring) which are covered later in this book may be applied in this context. Scoring methods can be normalized across different mediums such as forums or print, and when done against a competing product, the results can be compelling.

Suppose you are a Nike employee and you want to know about how consumers are viewing the Nike Men's Roshe Run Shoes. The text mining steps to follow are:

1) **Define the problem and specific goals.** Using online reviews, identify overall positive or negative views. For negative reviews, identify a consistent cause of the poor review to be shared with the product manager and manufacturing personnel.

2) **Identify the text that needs to be collected.** There are running websites providing expert reviews, but since the shoes are mass market, a larger collection of general use reviews would be preferable. New additions come out annually, so old reviews may not be relevant to the current release. Thus, a shopping website like Amazon could provide hundreds of reviews, and since there is a timestamp on each review, the text can be limited to a particular timeframe.

3) **Organize the text.** Even though Amazon reviewers rate products with a number of stars, reviews with three or fewer stars may yield opportunities to improve. Web scraping all reviews into a simple csv with a review per row and the corresponding timestamp and number of stars in the next columns will allow the analysis to subset the corpus by these added dimensions.

4) **Extract features.** Reviews will need to be cleaned so that text features can be analyzed. For this simple example, this may mean removing common words with little benefit like "shoe" or "nike," running a spellcheck and making all text lowercase.

5) **Analyze.** A very simple way to analyze clean text, discussed in an early chapter, is to scan for a specific group of keywords. The text-mining analyst may want to scan for words given their subject matter expertise. Since the analysis is about shoe problems one could scan for "fit," "rip" or "tear," "narrow," "wide," "sole," or any other possible quality problem from reviews. Then summing each could provide an indication of the most problematic feature. Keep in mind that this is an extremely simple example and the chapters build in complexity and analytical rigor beyond this illustration.

6) **Reach an insight or recommendation.** Armed with this frequency analysis, a text miner could present findings to the product manager and manufacturing personnel that the top consumer issue could be "narrow" and "fit." In practical application, it is best to offer more methodologies beyond keyword frequency, as support for a finding.

1.6 A Real World Use Case

It is regularly the case that marketers learn best practices from each other. Unlike in other professions many marketing efforts are available outside of the enterprise, and competitors can see the efforts easily. As a result, competitive intelligence in this space is rampant. It is also another reason why novel ideas are often copied and reused, and then the novel idea quickly loses salience with its intended audience. Text mining offers a quick way to understand the basics of a competitor's text-based public efforts.

When I worked at amazon.com, creating the social customer service team, we were obsessed with how others were doing it. We regularly read and reviewed other companies' replies and learned from their missteps. This was early 2012, so customer service in social media was considered an emerging practice, let alone being at one of the largest retailers in the world. At the time, the belief was that it was fraught with risk. Amazon's legal counsel, channel marketers in charge of branding and even customer service leadership were weary of publically acknowledging any shortcomings or service issues. The legal department was involved to understand if we were going to set undeliverable expectations or cause any tax implications on a state-by-state basis. Further, each brand owner, such as Amazon Prime, Amazon Mom, Amazon MP3, Amazon Video on Demand, and Amazon Kindle had cultivated their own style of communicating through their social media properties. Lastly, customer service leadership had made multiple promises that reached all the way to Jeff Bezos, the CEO, about flawless execution and servicing in this channel demonstrating customer centricity. The mandate was clear: proceed, but do so cautiously and do not expand faster than could be reasonably handled to maintain quality set by all these internal parties. The initial channels we covered were the two "Help" forums on the site, then retail and Kindle Facebook pages, and lastly, Twitter. We had our own missteps. I remember the email from Jeff that came down through the ranks with a simple "?" concerning an inappropriate briefly posted video to the Facebook wall. That told me our efforts were constantly under review and that we had to be as good as or better than other companies.

Text mining proved to be an important part of the research that was done to understand how others were doing social media customer service. We had to grasp simple items like length of a reply by channel, basic language used, typical agent workload, and if adding similar links repeatedly made sense. My initial thought was that it was redundant to repeatedly post the same link, for example to our "contact us", form. Further, we didn't know what types of help links were best to post. Should they be informative pages or forms or links to outside resources? We did not even know how many people should be on the team and what an average workload for a customer service representative was.

In short, the questions basic text mining can help with are

1) What is the average length of a social customer service reply?
2) What links were referenced most often?
3) How many people should be on the team? How many social replies is reasonable for a customer service representative to handle?

Channel by channel we would find text of some companies already providing public support. We would identify and analyze attributes that would help us answer these questions. In the next chapter, covering basic text mining, we will actually answer these questions on real customer service tweets and go through the six-step process to do so.

Looking back, the answers to these questions seem common sense, but that is after running that team for a year. Now social media customer service has expanded to be the norm. In 2012, we were creating something new at a Fortune 50 fast growing company with many opinions on the matter, including "do not bother!" At the time, I considered Wal-Mart, Dell and Delta Airlines to be best in class social customer service. Basic text mining allowed me to review their respective replies in an automated fashion. We spoke with peers at Expedia but it proved more helpful to perform basic text mining and read a small sample of replies to help answer our questions.

1.7 Summary

In this chapter you learned

- the basic definition of practical text mining
- why text mining is important to the modern enterprise
- examples of text mining used in enterprise
- the challenges facing text mining
- an example workflow for processing natural language in analytical contexts
- a simple text mining example
- when text mining is appropriate

2

Basics of Text Mining

In this chapter, you'll learn

- how to answer the basic social media competitive intelligence questions in Chapter 1's case study
- what the average length of a social customer service reply is
- what links were referenced most often
- how many people should be on a social media customer service team and how many social replies are reasonable for a customer service representative to handle
- what are the two approaches to text mining and how they differ
- common Base R functions and specialized packages for string manipulation

2.1 What is Text Mining in a Practical Sense?

There are technical definitions of text mining all over the Internet and in academic books. The short definition in Chapter 1, ("the process of distilling actionable insights from text") alludes to a practical application rather than relating to idle curiosity. As a practitioner, I prefer to think about the definition in terms of the value that text mining can bring to an enterprise. In Chapter 1 we covered a definition of text mining and expanded on its uses in a business context. However, in more approachable terms an expanded definition might be:

> Text mining represents the ability to take large amounts of unstructured language and quickly extract useful and novel insights that can affect stakeholder decision-making.

Text mining does all this without forcing an individual to read the entire corpus (pl. corpora). A graphical representation of the perspective given in Chapter 1 is shown in Figure 2.1. In the social customer service case the problem was

Text Mining in Practice with R, First Edition. Ted Kwartler.

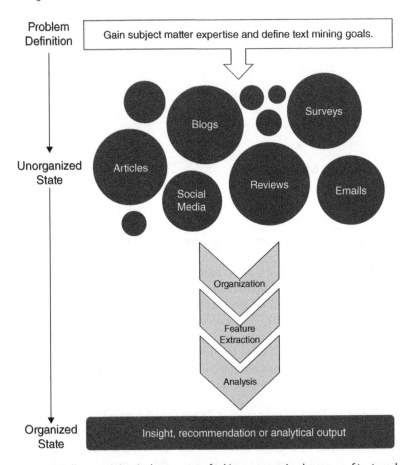

Figure 2.1 Recall, text mining is the process of taking unorganized sources of text, and applying standardized analytical steps, resulting in a concise insight or recommendation. Essentially, it means going from an unorganized state to a summarized and structured state.

reasonably well defined in order to inform operational decisions. The figure is a review of the mental map for transitioning from a defined problem and an unorganized state of data to an organized state containing the insight.

The main point of this practical text mining definition is that it views text mining as a means to an end. Often there are "interesting" text analyses that are performed but that have no real impact. If the effort does not confirm existing business instincts or inform new ones, then the analysis is without merit beyond the purely technical. An example of non-impactful text mining occurred when a vendor tried to sell me on the idea of sentiment analysis scoring for customer satisfaction surveys. The customer ranked the service interaction as "poor" or "good" in the first question of the survey. Running sentiment analysis on a subset of the "poor" interactions resulted in "negative" sentiment for all the survey notes. But confirming that poor interactions had negative sentiment in a later text-based question in no way helped to improve customer service operations. The customer should be trusted in question one! Companies delivering this type of nonsense are exactly the reason that text mining has never fully delivered on its expected impact.

The last major point in the definition is that the analyst should not need to read the entire corpus. Further, having multiple reviewers of a corpus doing text analytics causes problems. From one reviewer to another I have found a widely disparate understanding of the analysis deliverable. Reviews are subjective and biased in their approach to any type of scoring or text analysis. The manual reviewers represent another audience and, as communication theory states, messages are perceived by the audience not the messenger. It is for this reason that I prefer training sets and subjectivity lexicons, where the author has defined the intended sentiment explicitly rather than having it performed by an outside observer. Thus, I do not recommend a crowd-sourcing approach to analysis, such as mturk or crowdflower. These services have some merit in a specialized context or limited use, but overall I find them to be relatively expensive for the benefit. In contrast, interpreting biases in methodology through a code audit, and reviewing repeatable steps leading to the outcome helps to provide a more consistent approach. I do recommend the text miner to read portions of the corpus to confirm the results but not to read the entire corpus for a manual analysis.

Your text mining efforts should strive to create an insight while not manually reading entire documents. Using R for text mining ensures that you have code that others can follow and makes the methods repeatable. This allows your code to be improved iteratively in order to try multiple approaches.

Despite the technological gains in text mining over the past five years, some significant challenges remain. At its heart, text is unstructured data and is often high volume. I hesitate to use "big data" because it is not always big, but it still can represent a large portion of an enterprise's data lake. Technologies like Spark's ML-Lib have started to address text volume and provide structuring methods at scale. Another remaining text mining concern that is part of the human condition is that text represents expression and is thereby impacted by individualistic expressions and audience perception. Language continues to evolve collectively and individually. In addition, cultural differences impact language use and word choice. In the end, text mining represents an attempt to hit a moving target as language evolves, but the target itself isn't clearly defined. For these reasons text mining remains one of the most challenging areas of data science and among the most fun to explore.

Where does text mining fit into a traditional data science machine learning workflow?

Traditionally there are three parts to a machine learning workflow. The initial input to the process is historical data, followed by a modeling approach and finally the scoring for both new observations and to provide answers. Often the workflow is considered circular because the predictions inform the problem definition, and the modeling methods used, and also the historical data itself will evolve over time. The goal of continuous feedback within a machine learning workflow is to improve accuracy.

The text mining process in this book maps nicely to the three main sections of the machine learning workflow. Text mining also needs historical data from which to base new outcomes or predictions. In the case of text the training data is called a corpus or corpora. Further, in both machine learning and text mining, it is necessary to identify and organize data sources.

The next stage of the machine learning workflow is modeling. In contrast to a typical machine learning algorithm, text mining analysis can encompass non-algorithmic reasoning. For example, simple frequency analysis can sometimes yield results. This is more usually linked to exploratory data analysis work than in a machine learning workflow. Nonetheless, algorithm modeling can be done in text mining and is covered later in this book.

The final stage of the machine learning workflow is prediction. In machine learning, this section applies the model to new data and can often provide answers. In a text mining context, not only can text mining based algorithms function exactly the same, but also this book's text mining workflow shows how to provide answers while avoiding "curiosity analysis."

In conclusion, data science's machine learning and text mining workflows are closely related. Many would correctly argue that text mining is another tool set in the overall field of data discovery and data science. As a result, text mining should be included within a data science project when appropriate and not considered a mutually exclusive endeavor.

2.2 Types of Text Mining: Bag of Words

Overall there are two types of text mining, one called "bag of words" and the other "syntactic parsing," each with its benefits and shortcomings. Most of this book deals with bag of words methods because they are easy to understand and analyze and even to perform machine learning on. However, a later chapter is devoted to syntactic parsing because it also has benefits.

Bag of words treats every word – or groups of words, called n-grams – as a unique feature of the document. Word order and grammatical word type are not captured in a bag of words analysis. One benefit of this approach is that it is generally not computationally expensive or overwhelmingly technical to organize the corpora for text mining. As a result, bag of words style analysis can often be done quickly. Further, bag of words fits nicely into machine leaning frameworks because it provides an organized matrix of observations and attributes. These are called document term matrices (DTM) or the transposition, term document matrices (TDM). In DTM, each row represents a document or individual corpus. The DTM columns are made of words or word groups. In the transposition (TDM), the word or word groups are the rows while the documents are the columns.

Don't be overwhelmed, it is actually pretty easy once you see it in action! To make this real, consider the following three tweets.

- *@hadleywickham:* "How do I hate thee stringsAsFactors=TRUE? Let me count the ways #rstats"
- *@recodavid:* "R the 6th most popular programming language in 2015 IEEE rankings #rstats"
- *@dtchimp:* "I wrote an #rstats script to download, prep, and merge @ACLEDINFO's historical and realtime data."

This small corpus of tweets could be organized into a DTM. An abbreviated version of the DTM is in Table 2.1.

The transposition of the DTM is the term document matrix (TDM). This means that each tweet is a column in the matrix and each word is a row. Within the bag of words text mining approach, the type of analysis dictates the type of matrix used. Table 2.2 is an abbreviated TDM of the small three-tweet corpus.

In these examples, the DTM and TDM are merely showing word counts. The matrix shows the sum of the words as they appeared for the specific tweet. With the organization done, you may notice that all mention #rstats. So without reading all the tweets, you could simply surmise, based on frequency, that the general topic of the tweets most often is somehow related to R. This simple frequency analysis would be more impressive if the corpus contained tens of thousands of tweets. Of course, there are many other ways to approach the matter and different weighting schemes used in these types of matrices.

Table 2.1 An abbreviated document term matrix, showing simple word counts contained in the three-tweet corpus.

Tweet	@ acledinfo's	#rstats	2015	6th	And	Count	Data	Download	...
1	0	1	0	0	0	1	0	0	...
2	0	1	1	1	0	0	0	0	...
3	1	1	0	0	2	0	1	1	...

Table 2.2 The term document matrix contains the same information as the document term matrix but is the transposition. The rows and columns have been switched.

Word	Tweet1	Tweet2	Tweet3
@acledinfo's	0	0	1
#rstats	1	1	1
2015	0	1	0
6th	0	1	0
And	0	0	2
Count	1	0	0
Data	0	0	1
Download	0	0	1
...	...	...	...

However, the example is sound and shows how this simple organization can start to yield a basic text mining insight.

2.2.1 Types of Text Mining: Syntactic Parsing

Syntactic parsing differs from bag of words in its complexity and approach. It is based on word syntax. At its root, syntax represents a set of rules that define the components of a sentence that then combine to form the sentence itself (similar to building blocks). Specifically, syntactic parsing uses part of speech (POS) tagging techniques to identify the words themselves in a grammatical or useful context. The POS step creates the building blocks that make up the sentence. Then the blocks, or data about the blocks, is analyzed to draw out the insight. The building block methodologies can become relatively complicated. For instance, a word can be identified as a noun "block" or more specifically as a proper noun "block." Then that proper noun tag or block can be linked to a verb and so on until the blocks add up to the larger sentence tag or block. This continues to build until you complete the entire document.

More generally, tagging or building block methodologies can identify sentences; the internal sentence components such as the noun or verb phrase; and even take an educated guess at more specific components of the sentence structure. Syntactic parsing can identify grammatical aspects of the words such as nouns, articles, verbs and adjectives. Then there are dependent part of speech tags denoting a verb linking to its dependent words such as modifiers. In effect, the dependent tags rely on the primary tags for basic grammar and sentence structure, while the dependent tag is captured as metadata about the original tag. Additionally, models have been built to perform sophisticated tasks including naming proper nouns, organizations, locations, or currency amounts. R has a package relying on the OpenNLP (open [source] natural language processing) project to accomplish these tasks. These various tags are captured attributes as metadata of the original sentence. Do not be overwhelmed; the simple sentence below and the accompanying Figure 2.2 will help make this sentence deconstruction more welcoming.

Consider the sentence: "Lebron James hit a tough shot."

When comparing the two methods, you should notice that the amount of information captured in a bag of words analysis is smaller. For bag of words, sentences have attributes assigned only by word tokenization such as single words, or two-word pairs. The frequencies of terms – or sometimes the inverse frequencies – are recorded in the matrix. In the above sentence that may mean having only single tokens to analyze. Using single word tokens, the DTM or TDM would have no more than six words. In contrast, syntactic parsing has many more attributes assigned to the sentence. Reviewing Figure 2.2, this sentence has multiple tags including sentence, noun phrase, verb phrase, named entity, verb, article, adjective and noun. In this introductory book, we spend most of our time using the bag of words methodology for the basis of our foundation, but there is a chapter devoted to R's openNLP package to demonstrate part of speech tagging.

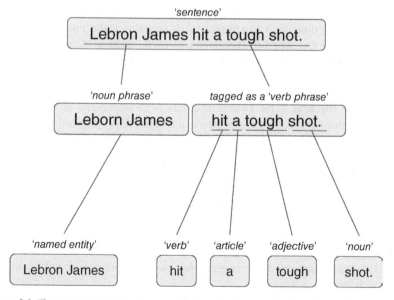

Figure 2.2 The sentence is parsed using simple part of speech tagging. The collected contextual data has been captured as tags, resulting in more information than the bag of words methodology captured.

2.3   The Text Mining Process in Context

1) **Define the problem and the specific goals.** Let's assume that we are trying to understand Delta Airline's customer service tweets. We need to launch a competitive team but know nothing about the domain or how expensive this customer service channel is. For now we need to answer these questions.
   a) What is the average length of a social customer service reply?
   b) What links were referenced most often?
   c) How many people should be on a social media customer service team? How many social replies are reasonable for a customer service representative to handle?

   Although the chapter covers more string manipulations beyond those needed to answer these questions, it is important to understand common string-related functions since your own text mining efforts will have different questions.

2) **Identify the text that needs to be collected.** This example analysis will be restricted to Twitter, but one could expand to online forums, Facebook walls, instagram feeds and other social media properties.

Getting the Data

Please navigate to www.tedkwartler.com and follow the download link. For this analysis, please download "oct_delta.csv". It contains Delta tweets from the Twitter API from October 1 to October 15, 2015. It has been cleaned up so we can focus on the specific tasks related to our questions.

---

3) **Organize the text.** The Twitter text has been organized already from a JSON object with many parameters to a smaller CSV with only tweets and date information. In a typical text mining exercise the practitioner will have to perform this step.
4) **Extract features.** This chapter is devoted to using basic string manipulation and introducing bag of words text cleaning functions. The features we extract are the results of these functions.
5) **Analyze.** Analyzing the function results from this chapter will lead us to the answers to our questions.
6) **Reach an insight or recommendation.** Once we answer our questions we will be more informed in creating our own competitive social customer service team.

   a) What is the average length of a social customer service reply? We will use a function called nchar to assess this.

b) What links were referenced most often? You can use `grep`, `grepl` and a summary function to answer this question.

c) How many people should be on a social media customer service team? How many social replies are reasonable for a customer service representative to handle? We can analyze the agent signatures and look at it as a time series to gain this insight.

2.4 String Manipulation: Number of Characters and Substitutions

At its heart, bag of words methodology text mining means taking character strings and manipulating them in order for the unstructured to become structured into the DTM or TDM matrices. Once that is performed other complex analyses can be enacted. Thus, it is important to learn fundamental string manipulation. Some of the most popular string manipulation functions are covered here but there are many more. You will find the `paste`, `paste0`, `grep`, `grepl` and `gsub` functions useful throughout this book.

R has many functions for string manipulation automatically installed within the base software version. In addition, the common libraries extending R's string functionality are `stringi` and `stringr`. These packages provide simple implementations for dealing with character strings.

To begin all the scripts in this book I create a header with information about the chapter and the purpose of the script. It is commented out and has no bearing on the analysis, but adding header information in your scripts will help you stay organized. I added my Twitter handle in case you, as the reader, need to ask questions. As your scripts grow in number and complexity having the description at the top will help you remember the purpose of the script. As the scripts you create are shared with others or inherited, it is important to have some sort of contact information such as an email so the original author can be contacted if needed. Additionally, as part of my standard header for any text mining, it is best to specify two system options that have historically been problematic. The first option states that strings are not to be considered factors. R's default understanding of text strings is to treat them as individual factors like "Monday," "Tuesday" and so on with distinct levels. For text mining, we are aggregating strings to distill meaning, so treating the strings as individual factors makes aggregation impossible. The second option is for setting a system locale. Setting the system location helps overcome errors associated with unusual characters not recognized by R's default locale. It does not fix all of them but I have found it helps significantly. Below is a basic script header that I use often in my text mining applications.

```
#Text Mining in R
#Ted Kwartler
#Twitter: @tkwartler
#Chapter 2 Text Mining Basics

#Set Options
options(stringsAsFactors=F)
Sys.setlocale('LC_ALL','C')
```

Next I bring in libraries to aid in our basic string exercises. In addition to base functions, I like to use `stringi`, `stringr` and `qdap` for these, although there are many other libraries that can aid in string manipulation.

```
library(stringi)
library(stringr)
library(qdap)
```

You can now jump into working with text for the first time, since you have a concise reference header, our specialized text libraries loaded, and some sample text similar to our Delta case study.

The base function `nchar` will return the number of characters in a string. The function is vectored, meaning that it can be applied to a character column directly without using an additional `apply` function. In contrast, some functions work only on a part of the data like an individual cell in a data frame. In practical application, `nchar` may be of interest if you are reviewing competitive marketing materials. The function can also be used in other functions to help clean up unusually short or even blank documents in a corpus. It is worth noting that `nchar` does count spaces as characters.

The code below references the first six rows of the corpus and the last column containing the text. Instead of using the `head` function you can reference the entire vector with `nchar` or you can look at any portion of the data frame by referencing the corresponding index position.

```
text.df<-read.csv('oct_delta.csv')
nchar(head(text.df$text))
```

The returned answer:

```
[1] 119 110 78 65 137 142
```

If you want to look at only a particular string, you can refer to it by its exact index location. Here apply `nchar` to only the fourth row and last column containing a single tweet.

```
nchar(text.df[4,5])
```

The individual answer returned:

```
[1] 65
```

Using this function you can easily answer your first case study question. The code below calculates the average for the number of characters for the entire vector tweets.

```
mean(nchar(text.df$text))
```

What is the average length of a social customer service reply? The answer is approximately 92 characters. Since tweets can be a maximum of 140 characters, the insight here is that agents are concise and not often maximizing the Twitter character limit. In the data set, there are cases of a long message being broken up into multiple tweets. In this type of analysis it is best to have more data and to subset these multiple tweets to ensure accuracy.

Another use for the nchar function is that it can be used to omit strings with a length equal to 0. This can help remove blank documents from a corpus, as is sometimes the case in a table with extra blank rows at the bottom. To do this you can use the subset function along with the nchar function as shown below. If you end up with blank documents, it may make analyzing the entire collection difficult. So this function keeps only documents with the number of characters greater than 0.

```
subset.doc<-subset(text.df,nchar(text.df$text)>0)
```

The following functions do not necessarily answer the questions in the case study but nonetheless are important to know. These functions replace defined patterns in strings. These functions represent a simple way to substitute parts of strings. As a result these useful functions are often used in unifying text and aggregating important terms. This family of functions can be useful to clean up unwanted punctuation, aggregate terms, change ordinal abbreviations (e.g. "2nd") or change acronyms to their long form. For example, if you are analyzing text with a lot of tweets, you may encounter a lot of "@" symbols which could be removed using these functions. Another example would be to aggregate terms. We could look for patterns of "Coke" and change them to "Coke a Cola" if it made sense in the problem context.

The first function is sub. The sub function looks for the first pattern match in a string and replaces it. In the data set, Delta's first customer service tweet is "@mjdout I know that can be frustrating..we hope to have you parked and deplaned shortly. Thanks for your patience. *AA." We can substitute one string for another using the code below.

```
sub('thanks','thank you',text.df[1,5], ignore.case=T)
```

The console prints the resulting tweet with changed words as shown below.

```
[1] "@mjdout I know that can be frustrating..we hope
to have you parked and deplaned shortly. thank you for
your patience. *AA"
```

The sub function is also vectored so that it can be applied to a column, and replacements will occur for each of the first pattern matches by row. The index below is applying sub to the first five tweets' rows within the fifth column but can just as easily be used on the entire column.

```
sub('pls','please',text.df[1:5,5], ignore.case=F)
```

This changes the 'pls' in tweets two, three and four. Three replacements were made because each instance is the first pattern for that specific row.

Next we will discuss the function gsub. This global substitution function will replace not just the first instance of a pattern but instead all instances. Consider the following example text string to illustrate the difference between sub and gsub.

```
fake.text<-'R text mining is good but text mining in
python is also'
sub('text mining','tm', fake.text, ignore.case=F)
gsub('text mining','tm', fake.text, ignore.case=F)
```

Using just sub the first pattern match of text mining is replaced with tm while the second one is not. However, using gsub allows the function to match more than once within a string. As a result the second result has both substitutions completed. As was the case with sub, the gsub function can be applied to an entire vector of strings.

```
[1] "R tm is good but tm in python is also"
```

The gsub function is useful for removing specific characters in entire corpora. In the fifth Delta tweet R has parsed the "&" ampersand to "&amp." We can remove this pattern entirely or substitute a new word by using gsub on a specific row.

```
gsub('&amp','',text.df[5,5])
```

When you look closely you see that the pattern "&amp" is being replaced with nothing inside the quotes as the second entry to the function. This is a simple way to drop the matched pattern. It can also be used, sometimes clumsily, to remove punctuation. To remove a single punctuation, the pattern to search for would merely be a punctuation like "!" or "@". However, to remove all punctuation using the regular expression and gsub you can call on the following code.

```
gsub('[[:punct:]]','',text.df[1:5,5])
```

Another point that is worth noting is that the last part of both sub and gsub function is the parameter ignore.case=F). This tells the function to explicitly match the pattern. Changing the F to T for true will tell the sub or gsub function that it can match either upper or lowercase strings.

The qdap package, which is a wonderfully expansive package providing a lot of useful text mining tools, offers a very convenient wrapper for gsub. It is called mgsub or multiple global substitutions. This allows an R programmer to pass a vector of pattern matches to be replaced with another vector. It is compact and makes repeating many multiple substitutions easy. To begin with, we create a string vector of patterns to be matched. Then we create another string vector of replacements. Lastly we invoke the mgsub function applied to the fake.text object created earlier. In the code below "good" will be replaced with "great," "also" will be replaced with "just as suitable," and "text mining" will be replaced by "tm."

```
library(qdap)
patterns<-c('good','also','text mining')
replacements<-c('great','just as suitable','tm')
mgsub(patterns,replacements,fake.text)
```

Rather than having three separate gsub function calls we are able to accomplish three substitutions more efficiently. The result is below.

```
[1] "R tm is great but tm in python is just as
suitable"
```

The mgsub function is the easiest and best way to programmatically do multiple substitutions. In text analysis an analyst may end up with a vector of words that were identified earlier and needs to be replaced, modified or aggregated. In my experience, mgsub is the best way forward instead of tedious individual gsub function calls or a custom function.

In practical business application the substitute functions are useful. We can change repeating words to aggregate terms. For example, changing acronyms is easy using sub functions. At Amazon, customer service agents refer to "WMS" in call notes. This stands for "where's my stuff" and represents a caller who is looking for a shipment. Using the sub function one can revert WMS to the entire phrase or vice versa.

Tip: Sometimes appropriate yet unintended consequences can occur. For example, at a large organization one of my scripts was replacing RT with a blank space to remove a retweet designation. Using gsub had the unintended consequence of changing words like "airport" to "airpo." So beware when using gsub as you may encounter unplanned consequences.

2.4.1 String Manipulations: Paste, Character Splits and Extractions

Another useful function, especially when dealing with multiple columns of text to be analyzed, is the paste function. For business analysts used to Excel, paste is the same as the concatenation function used for vectors. In the example data set and case study, you are trying to understand the number of tweets handled by an agent for a specific timeframe. As a result we need to paste the month, date and year columns from the data frame.

First you need to change the month abbreviations to the corresponding number. To do this you can simply use a substitute function from the previous section. If you were executing this analysis for real, you would be looking at more than a single month in the data set, so the code below subs all 12 months of the year.

```
patterns<-c('Jan','Feb','Mar','Apr','May','Jun','Jul',
'Aug','Sep','Oct','Nov','Dec')
replacements<-seq(1:12)
text.df$month<-mgsub(patterns,replacements,text.
df$month)
text.df$combined<-paste(text.df$month,
text.df$date,text.df$year, sep='-')
```

This creates a new column or vector called "combined." The function takes the three string vectors and combines them with a separator. Another variant of the paste function is paste0. The paste0 function can be used if there are no separating characters needed and is therefore slightly more efficient in some cases.

If you were completing this analysis for real and needed to understand the agent's tweet patterns as a time series, you would need to change the text.df$combined vector to dates. The lubridate package is then used in order to switch the newly created dates into an official date format. Once this is done, an analyst can use all date-related functions (like difference times) to explore work load arrival patterns.

```
library(lubridate)
text.df$combined<-mdy(text.df$combined)
```

With the date cleaned up and having covered some basic string manipulation functions, you now turn your attention to actually understanding the agent workload. Another base R function that acts similarly is strsplit. The strsplit function creates subset strings by matching character patterns.

In setting up Amazon's customer service team, I reviewed other companies' tweets similar to these, in an effort to learn the number of agents that other companies were using and also how many tweets each agent could handle in a normal shift. This could aid in proper benchmarking and workforce management. Reviewing the first two example tweets from the data frame you can see that agents are adding personal initials to each tweet.

```
text.df$text[1:2]
[1] "@mjdout I know that can be frustrating..we hope
to have you parked and deplaned shortly. Thanks for
your patience.  *AA"
```

```
[2] "@rmarkerm Terribly sorry for the inconvenience.
If we can be of assistance at this time, pls let us
know.  *AA"
```
String split on the asterisk to identify the agent for each tweet.
```
agents<-strsplit(text.df$text,'[*]')
```

Notice the pattern is [*] not just *. This is because a standalone * in regular expressions is considered a wildcard, meaning match anything. In this example, that would not be useful. However, adding the brackets around the asterisk tells the strsplit function to treat the asterisk as simply the character. The result is a list with each second value holding the information we are interested in. In practical application, doing this on many weeks' worth of tweets allows a text miner to aggregate the customer service agent initials, and summarize along with the timestamp of the tweet in order to deduce how that specific team performs. Here, the same agent "AA" signed both tweets.

```
agents[1:2]
[[1]]
[1] "@mjdout I know that can be frustrating..we hope
to have you parked and deplaned shortly. Thanks for
your patience.  "
[2] "AA"
[[2]]
[1] "@rmarkerm Terribly sorry for the inconvenience.
If we can be of assistance at this time, pls let us
know.  "
[2] "AA"
```

The strsplit function works as long as all agents are using the same pattern to close their messages. If an agent uses another character such as a dash instead of an asterisk, then the strsplit function would miss that tweet signature. It may be the case that agents use a mixture of patterns to close messages. Thus a custom function may be better to accomplish this or a similar task where the text miner needs to capture the final two characters from a document.

An example of a custom function follows. In this case, it is called "last.chars." You need to specify a piece of text and a number when invoking the function. The function will return the object called 'last' that represents the last number of letters in the string. The last.chars function works by using the substring function along with nchar. Substring extracts parts of a string based on a beginning and ending number. A quick example is shown below. The function extracts the portion of the overall string "R text mining is great" beginning at the 18th character and ending at the 22nd. The function counts spaces in this withdrawal, and the result is only the characters in the word "great," as shown.

```
substring('R text mining is great',18,22)
[1] "great"
```

The last.chars function merely creates the numbers for substring dynamically. The first numerical substring function input uses the number of characters in the string minus the number given when calling the last. chars function plus one. The second input represents the number to end the extraction. Since the function is meant to capture the end of the string, the second number is the total number of characters in the string itself, meaning grab characters until the end is reached.

```
last.chars<-function(text,num){
    last<-substr(text, nchar(text)-num+1,nchar(text))
    return(last)
}
```

Applying the last.chars function to the "R text mining is great" function is straightforward and can capture the same information as substring. The advantage is that last.chars is dynamic and does not extract based on a specific character like strsplit.

```
last.chars('R text mining is great',5)
[1] "great"
```

Now you need to apply the last.chars function to the vector along with the number two. This will dynamically grab the last two characters of the vector. This solves the problem of splitting on an exact character match in case the agent used a different character to signify the start of the signature.

```
last.chars(text.df$text[1:2],2)
[1] "AA" "AA"
```

Armed with your custom function, you can use it on one or more weeks of @ deltaassist tweets. For the sake of learning, subset based on your pasted and cleaned dates and then perform an analysis to see the hardest working Delta customer service agent. In the next code, you create an object called "weekdays." It represents a subset of the entire data frame in between October 5 and October 9. Then in order to make sense of them for analysis, you need to treat them as categorical factors, so you can call on the table function.

Table 2.3 @DeltaAssist agent workload – *The abbreviated table demonstrates a simple text mining analysis that can help with competitive intelligence and benchmarking for customer service workloads.*

Agent	AA	AD	AN	BB	CK	CM	DD	DR	EC	HW	...	VI	VM	WG
Cases	32	7	6	14	7	4	5	6	5	5	...	8	9	35
Agent	AA	AD	AN	BB	CK	CM	DD	DR	EC	HW	...	VI	VM	WG
Cases	32	7	6	14	7	4	5	6	5	5	...	8	9	35

```
weekdays<-subset(text.df,text.df$combined >= mdy('10-
05-2015') & text.df$combined<= mdy('10-09-2015'))
table(as.factor(last.chars(weekdays$text,2)))
```

The result is a table summary of the last two characters for each tweet within the time period. Some tweets are continuations of customer service cases and are showing up as "/2" or something similar. Because these are continuations, and we are concerned about individual caseloads, to answer our earlier question you can safely ignore them, although an interesting analysis may be to later understand how often agents need to create these truncated messages. Table 2.3 is an abbreviated version of the `table` output. You can quickly determine that the busiest agent is WG, and with a little more effort you can examine the average among all agents.

Using the above code answers the simple case study question about agent workload in a week. You can easily extend this analysis by looking at days of the week and actual times of day for responses. In fact, Twitter's api returns the entire timestamp of tweets with time of day. With that you could extend this agent workload analysis to make an educated guess at each agent's work shift.

2.5   Keyword Scanning

The functions `grep` and `grepl` have a long history of use in computer programming. R has inherited these commands from Unix, created over 40 years ago! In fact, the `grep` commands are so often used that the command `grep` is both a noun and a verb. While `grep` and `grepl` sound like alien terms, the commands are in fact merely searching for a regular expression pattern. More specifically, the functions stands for "global regular expression print."

The pseudo code for both is straightforward. The "l" within `grepl` changes the output printed, but the function parameters are the same. The pseudo code shown next.

```
Grep-Text-search(character pattern to search, where
should the search happen?, should uppercase matter or
be ignored?)
```

Simply calling `grep` will return the position of the searched pattern. For example, if the second document of a corpus has the string pattern "text," then `grep` will return a [2]. In contrast, adding the lowercase l, "l," to `grep` will return a logical vector of TRUE or FALSE for every place it searched. TRUE means the pattern that was searched appeared at least once while FALSE means it was not found. It is important to note that `grep` does not count the number of times the search term appeared, only whether it appeared at least once.

To start searching for terms you can pass a pattern to the `grep` function. Here you are looking for the pattern "sorry" within all DeltaAssist tweets. We are explicitly telling the `grep` command to look within the column called "text". Lastly, you are telling `grep` to ignore the case of the pattern because the last parameter is set to T or TRUE.

```
grep('sorry', text.df$text,ignore.case=T)
```

Grep will return the position of the tweets that contain this pattern at least once. After typing the command above into your console, R will return a vector of numbers including 2, 18, 22 and 26. These represent the tweet position in the original data frame that contains the character pattern "sorry." Look at one of these tweets.

```
text.df$text[413]
[1] "@RSstudio ...you on your way as soon as we can. Sorry
for the wait. *SB 2/2"
```

The `grepl` command functions similarly but the returned information is different. Here the returned information is a logical vector. For the DeltaAssist tweets, there are 1377 different tweets, one per row. As a result `grepl` will return a vector of 1377 true or false results when searching for the specified pattern. Since it is a logical return, R can also treat True as 1 and False as 0. This can come in handy when trying to summarize the occurrence of a word. The code below illustrates how to calculate the percentage of the time Delta customer service agents state that they are sorry from among all tweets in the timeframe. First, you create an object called `sorry` using `grepl`. If you then type `sorry` into the console, you see 1377 TRUE or FALSE returns, one for each row. Although the sorry vector is True or False, R will treat True as 1 and False as zero so you can sum it. We can simply sum the true and divide by the number of tweets in the entire data frame. The result is that 0.131 or 13.1% of all tweets contain at least a sorry. At Amazon, our legal department was leery of over apologizing and perhaps taking binding public acceptance of blame. It was through analyses like this that we argued that apologizing is part of social customer service expectations and therefore does not pose a large risk.

```
sorry<-grepl('sorry', text.df$text,ignore.case=T)
sum(sorry)/nrow(text.df)
```

Next you may want to search for more than one term at a time. You can do this by passing more than one word to the grep functions. However, since you are working with regular expressions, you must combine them in a specific and logical manner. The "pipe" or straight line is located above the enter key on some keyboards and above the left-hand Ctrl key on others. The pipe represents an "or" in between string patterns.

In this code, you are looking for any tweets that contain "sorry" or "apologize". Within the function section that holds the character patterns, you are combining a vector of words with the pipe in between them. If you wanted to add more words with an "or" relationship, you can simply add another pipe and the new pattern to search for. The "|" represents the "alternative operator" for regular expressions.

```
grep(c('sorry|apologize'),text.df$text,ignore.case=T)
```

To answer the final case study question you can identify the tweets containing a link, and you can also compare it to how often the agents are sharing a phone number. You need to deconstruct the regular expression below to help understand it. The first pattern you are looking for is anything that has "http" since urls often start with the hypertext transfer protocol pattern "http." The next pattern you are looking for may yield some false positives but is likely good enough for this basic analysis. The pattern is any three digits appearing in a row or any four digits appearing in a row. This is because phone numbers in the US follow a predictable xxx-xxx-xxxx pattern. In this expression, you assume that you are looking for either of the numerical blocks of a phone number. This could cause problems if agents are using, say, a confirmation number that matches this pattern. However, agents do not usually share personal information like this over a public channel. In both cases, R is summing the TRUE from grep and dividing by the number of tweets in the data frame.

```
sum(grepl('http', text.df$text, ignore.case =
T))/nrow(text.df)
sum(grepl('[0-9]{3})|[0-9]{4}', text.df$text))/
nrow(text.df)
```

The surprising insight here is that the phone numbers are twice as likely as the links to be shared (0.098 to 0.042). At Amazon, the preference is for self-service to an issue because it is best for the customer's time and cheaper for Amazon. So, in that case, Amazon agents contrast with Delta's because Amazon agents defer to a web page with the answer rather than instruct a customer to call.

To understand the links themselves that Delta agents are using, you can simply use the grep function to identify tweets with the "http" pattern and then apply a frequency analysis.

2.6 String Packages stringr and stringi

Earlier you learned how to identify if a particular character string or word is present at least once in a document. For this purpose, grep and grepl work well. However, if you are looking to count the number of times within a document that a string is found, rather than just its existence, then you will need to use a customized library. The stringi library provides a function for just this task. The stri_count function performs this task by returning a vector of 1377 numbers, one per tweet, which is the frequency count for the exact search term. Here you are searching for any instance of "http." In doing so you are able to identify the tweets that contain a url link. As part of the analysis that was done at Amazon, we needed to find out the common links to send people to, on social media. The functions in this chapter provided an easy way to scan and find messages containing links. This helped identify whether the links were to an input form or general information. In this case, it looks like DeltaAssist uses links sparingly in favor of direct messages through the Twitter platform. Changing the "http" to "DM," which is an abbreviation for a direct message, will scan for these customer service replies for comparison.

```
library(stringi)
stri_count(text.df$text, fixed='http')
```

Tip: The function works in a different order from grep or grepl. First you tell stri_count where to search and then you pass the pattern.

In the last section, you learned how to search for terms using grep and even how to search for multiple instances using the pipe, representing "or" between patterns. There are instances in which you may want to search for patterns using an "and" instead. To do so you need to load another useful package called stringr. Here you can stack the returned values with an ampersand to represent the fact that you need both searches to return a true result.

The code below returns a logical vector looking for the character pattern "http" within your tweets.

```
library(stringr)
str_detect(text.df$text, 'http')
```

In order to stack one of these logical statements, you can use the following code. This code is looking for http and DM and returns a TRUE if both patterns are present. In this case, it looks like DeltaAssist never asks for a DM and provides a URL link. When setting up a similar team, this type of information can be useful for creating operational procedures for customer service agents.

```
patterns <- with(text.df, str_detect(text.df$text,
'http') & str_detect(text.df$text, 'DM'))

text.df[patterns,5]
```

While this may look confusing, you can deconstruct it to make more sense. First you create an object called `patterns`. This is using two `str_detect` functions stacked with an ampersand. If you want to add more search patterns, then you would add another ampersand and `str_detect` function before the last parentheses. If you remember indexing a data frame, then the next line of code should make some intuitive sense. The `text.df` data frame is being indexed using the newly created patterns object but only the fifth column is returned. In this line of code we are removing all other information in the data frame to return just the text that matches both logical checks in the patterns object. This type of language exploration can be useful, as you may guess that Delta would be providing a link with helpful information and then would be asking for a DM, should the customer have any more questions. This type of text mining requires some subject matter expertise to specify the patterns to search for and is therefore limited but nonetheless useful.

Both `stringr` and `stringi` have many string manipulation functions and are worth exploring. For example, returning words by position in a sentence or making characters all upper or lowercase can be useful. From a text mining perspective the functions covered in this chapter should lay a solid foundation as you build your expertise. As you progress in skill level or are confronted with a specific use case beyond the general one outlined thus far, it would make sense to further explore these two important libraries.

2.7 Preprocessing Steps for Bag of Words Text Mining

Now that you have learned basic string manipulation you can expand to more interesting text mining. It is important to master the preprocessing steps and how to apply them. These preprocessing steps are consistent and foundational to most of the scripts in this book, no matter the analysis being performed or visual output. In Figure 2.1, the chevron style arrows in between the unstructured state and the insight represent the preprocessing steps and analysis that will be performed by R. Specifically the "organization" chevron is meant to encompass not only the collecting of text but also these preprocessing or cleaning steps. It should be noted that you can create custom preprocessing steps, depending on the analysis. For instance, in Twitter you may want to preprocess specific tokens such as "RT" or "#" by either removing retweets or explicitly identifying hash tokens as providing more context in the analysis. The cleaning steps outlined here represent common and foundational steps.

After setting the options in R that support text mining, you will load applicable libraries. These set R to realize that strings are not categorical variables and broaden the system location to avoid some encoding problems. For this exercise we will continue using the DeltaAssist tweets in object "text.df."

```
options(stringsAsFactors = FALSE)
Sys.setlocale('LC_ALL','C')
library(tm)
library(stringi)
```

Please remember when you read in files, create new objects or perform analysis that in R objects are held in RAM instead of on a hard drive. As a result, your computer's RAM can become a constraint on the amount of text and analysis that can be done. This example is merely 1377 tweets, so most modern laptops will be fine. However, as you explore larger corpora you will need to start increasing RAM, removing objects from the work stream that are no longer needed, or exploring packages such as SOAR and data.table. In fact, it is such a problematic issue that a many blogs deal with setting up cloud instances, which is a cheap alternative to buying a workstation with significant RAM.

You will need to keep track of the tweets. Since this data frame does not have unique IDs, you will create them using the code below in a new data frame called tweets.

```
tweets<-
data.frame(ID=seq(1:nrow(text.df)),text=text.df$text)
```

Now we must begin the text-cleaning task. The most common tasks we will perform in this book are lowering text, removing punctuation, stripping extra whitespace, removing numbers and removing "stopwords." Stopwords are common words that often do not provide any additional insight, such as articles. Table 2.4 describes each of the cleaning functions and provides an example.

You should note that the specific cleaning steps and transformations vary with the type of analysis. For instance, making all text lowercase makes finding proper nouns or "named entities" difficult. Removing numbers makes extracting dollar amounts impossible. Table 2.4 is foundational but not used in every text mining application.

Table 2.4 Common text-preprocessing functions from R's tm package with an example of the transformation's impact.

TM Function	Description	Before	After
tolower	Makes all text lowercase	Starbuck's is from Seattle.	starbuck's is from seattle.
removePunctuation	Removes punctuation like periods and exclamation points.	Watch out! That coffee is going to spill!	Watch out That coffee is going to spill
stripWhitespace	Removes tabs, extra spaces	I like coffee.	I like coffee.
removeNumbers	Removes numerals	I drank 4 cups of coffee 2 days ago.	I drank cups of coffee days ago.
removeWords	Removes specific words (e.g. he and & she) defined by the data scientists	The coffee house and barista he visited were nice, she said hello.	The coffee house barista visited nice, said hello.
stemDocument	Reduces prefixes and suffixes on words making term aggregation easier.	Transforming words to do text mining applications is often needed.	Transform word to do text mine applic is often needed.

The removeWords function requires another parameter, listing the exact words you want to remove. In the table example, I merely chose the common English stopwords – he, and, she – for removal. Table 2.5 contains all of the standard English stopwords used for the tm package. In order to customize this list, you can add or subtract words as needed. For example, if you were text mining legal documents you might want to customize the stopwords list by adding words like "defendant" and "plaintiff," since they will appear often in that context. The exact code to add to the stopwords list is covered later in this chapter. Table 2.5 lists the standard English stopwords.

Word	Word	Word	Word	Word	Word
i	by	they're	his	then	a
yours	after	they'd	them	both	while
herself	off	weren't	these	not	through
which	where	shouldn't	have	ourselves	in
was	other	there's	could	she	here
does	than	if	you've	theirs	few
she's	my	with	you'll	am	own
he'd	yourselves	below	haven't	had	your
isn't	its	under	cannot	i'm	hers
won't	whom	how	where's	they've	what

(Continued)

who's	be	such	because	she'll	are
the	doing	very	against	doesn't	do
at	we're	we	from	mustn't	he's
before	we'd	him	further	how's	you'd
on	wasn't	they	any	until	they'll
when	shan't	that	nor	into	didn't
most	here's	being	ours	down	that's
so	but	should	himself	once	an
me	for	i've	their	each	of
yourself	above	i'll	those	only	during
it	over	hasn't	has	you	out
who	why	can't	ought	her	there
were	some	when's	we've	themselves	more
did	too	or	he'll	is	same
it's	myself	about	hadn't	having	
she'd	he	to	couldn't	you're	
aren't	itself	again	why's	i'd	
wouldn't	this	all	as	we'll	
what's	been	no	between	don't	
and	would	our	up	let's	

Word	Word	Word	Word	Word	Word
i	by	they're	his	then	a
yours	after	they'd	them	both	while
herself	off	weren't	these	not	through
which	where	shouldn't	have	ourselves	in
was	other	there's	could	she	here
does	than	if	you've	theirs	few
she's	my	with	you'll	am	own
he'd	yourselves	below	haven't	had	your
isn't	its	under	cannot	i'm	hers
won't	whom	how	where's	they've	what
who's	be	such	because	she'll	are
the	doing	very	against	doesn't	do
at	we're	we	from	mustn't	he's
before	we'd	him	further	how's	you'd
on	wasn't	they	any	until	they'll
when	shan't	that	nor	into	didn't
most	here's	being	ours	down	that's
so	but	should	himself	once	an
me	for	i've	their	each	of
yourself	above	i'll	those	only	during
it	over	hasn't	has	you	out
who	why	can't	ought	her	there
were	some	when's	we've	themselves	more
did	too	or	he'll	is	same
it's	myself	about	hadn't	having	
she'd	he	to	couldn't	you're	
aren't	itself	again	why's	i'd	
wouldn't	this	all	as	we'll	
what's	been	no	between	don't	
and	would	our	up	let's	

Table 2.5 In common English writing, these words appear frequently but offer little insight. As a result, they are often removed to prepare a document for text mining.

Word	Word	Word	Word	Word	Word
i	by	they're	his	then	a
yours	after	they'd	them	both	while
herself	off	weren't	these	not	through
which	where	shouldn't	have	ourselves	in
was	other	there's	could	she	here
does	than	if	you've	theirs	few
she's	my	with	you'll	am	own
he'd	yourselves	below	haven't	had	your
isn't	its	under	cannot	i'm	hers
won't	whom	how	where's	they've	what
who's	be	such	because	she'll	are
the	doing	very	against	doesn't	do
at	we're	we	from	mustn't	he's
before	we'd	him	further	how's	you'd
on	wasn't	they	any	until	they'll
when	shan't	that	nor	into	didn't
most	here's	being	ours	down	that's
so	but	should	himself	once	an
me	for	i've	their	each	of
yourself	above	i'll	those	only	during
it	over	hasn't	has	you	out
who	why	can't	ought	her	there
were	some	when's	we've	themselves	more
did	too	or	he'll	is	same
it's	myself	about	hadn't	having	
she'd	he	to	couldn't	you're	
aren't	itself	again	why's	i'd	
wouldn't	this	all	as	we'll	
what's	been	no	between	don't	
and	would	our	up	let's	

It is worth noting that the transformation `stemDocument` may result in non-English words. When stemming documents, you may end up creating word fragments like "applic," so another transformation called `stem` `completion` may be needed afterwards to revert the fragments back to the most common complete word. For example "liking," "liked" and "like" would all be stemmed to "lik" then `stem completed` to "like."

The following code neatly organizes these foundational text transformations into functions. This makes applying them to various corpora easier and saves typing them repeatedly.

The first function is a wrapper for the base R `tolower` function. People online have noted that `tolower` fails when it encounters a special character that it is unable to recognize. Using base R's `tryCatch` function allows the function to ignore the error, keeping it from failing. The `tryCatch` function provides a way of handling unusual conditions that result in errors or warnings. In this case, we create another function called `tryTolower` that simply returns NA instead of failing.

```
# Return NA instead of tolower error
tryTolower <- function(x){
  # return NA when there is an error
  y = NA
  # tryCatch error
  try_error = tryCatch(tolower(x), error = function(e) e)
  # if not an error
  if (!inherits(try_error, 'error'))
    y = tolower(x)
  return(y)
}
```

Next you will define our stopwords. The code below creates a vector of word strings that includes the common English stopwords from above and also allows for adding to it. You can change the custom words by including quotes, and commas as shown for the custom words "lol," "smh" and "delta." This new character vector of words will be used later in our next transformation function. Twitter speak is evolving and often abbreviated. Here lol and smh are removed representing laughing out loud and shaking my head. This serves as an example of removing words based on the context of the corpora. The last custom word is delta. This is because the entire data frame is made of delta assist tweets so the word may be frequent yet not yield a new insight.

```
custom.stopwords <- c(stopwords('english'), 'lol',
'smh', 'delta')
```

Tip: It's best to rerun a text mining analysis with different custom stopwords to explore the conclusions that can be extracted. The words in the vector have to be lower case in order to be recognized and removed.

Next you will include the new `tryTolower` function as part of a larger pre-processing function. Here you create a function called `clean.corpus`. Within this function you can see specific foundational cleaning functions `removePunctuation`, `stripWhitespace`, `removeNumbers`, `tryTolower`, and `removeWords`. The custom `clean.corpus` function is passed a corpus object and then the corpus is repeatedly transformed with the specific preprocessing step. Note you use "tm_map" which is an interface function for transforming entire corpora. The corpus object is moved from step to step and rewritten as it moves through the function. For the newly created `tryTolower` function you have to add the additional "content_transformer" because you are using a customized version of "tolower" which modifies the content of an R object.

```
clean.corpus<-function(corpus){
  corpus <- tm_map(corpus,
content_transformer(tryTolower))
  corpus <- tm_map(corpus, removeWords,
custom.stopwords)
  corpus <- tm_map(corpus, removePunctuation)
  corpus <- tm_map(corpus, stripWhitespace)
  corpus <- tm_map(corpus, removeNumbers)
  return(corpus)
}
```

Remember as you customized the `clean.corpus` function, the order of operations is purposeful. We must lowercase the corpus to match the vector of lowercase stopword strings to be removed. Additionally the stopwords vector includes contractions like she'd, and isn't. Thus the removal of these stopwords needs to come before `removePunctuation`. Similarly once the apostrophes are removed, as part of the previous step, sometimes there is an additional white space that needs to be cleaned up as well using `stripWhitespace`.

Before applying these cleaning functions, you need to define the `tweets` object as your corpus or collection of natural language documents. Additionally you are preserving the metadata about each document. In this example, the unique information is the unique tweet ID you created. In this way, the ID and value become tag value pairs. This can aid you later, in other types of analysis, where the specific author is important, such when as measuring sentiment difference between authors. In order to preserve the metadata about the author or tweet, you need to create a specific mapping for R. This custom mapping tells R to interpret not only the text in question but also the associated information. In this case, it is the assigned twitter ID.

```
meta.data.reader <- readTabular(mapping=list(content=
'text', id='ID'))
```

With the customized functions in place you must tell R that you should treat the CSV containing 1377 tweets as a corpus or collection of documents. First you create an object called `corpus` by invoking the volatile `corpus` function and passing the tweets data frame to it. You explicitly tell R to use the "text" column as the text analysis vector. The ID column is still captured, so you may capture some metadata about the corpus.

```
corpus <- VCorpus(DataframeSource(tweets),
readerControl=list(reader=meta.data.reader))
```

Tip: Notice that you are creating a VCorpus. This particular type of corpus is a volatile corpus. This means that it is held in RAM memory in your computer. If you close R, shut down your computer or run out of power without saving, the VCorpus is lost, hence the volatility. In contrast a PCorpus creates a permanent corpus, because it is saved to permanent storage such as a hard drive.

With all the preprocessing transformation functions organized you must now apply them to the DeltaAssist corpus. Since you created a simple succinct `clean.corpus` function, you do not need to type them again for each corpus that you want to clean. This will be valuable later, when you are comparing multiple corpora in the same analysis.

```
corpus<-clean.corpus(corpus)
```

Tip: Depending on your operating system and version of R, you may encounter an error during the `clean.corpus` step. If your code states a "core" issue then change the `tm_map` functions to include the number of cores as shown here:

```
tm_map(corpus, removeNumbers, mc.cores=1)
```

One way to see information about the corpus object is to look at the list of documents. Here you examine the first document within the corpus. The first entry is captured as a plain text document along with some other basic information related to that particular document, or tweet in our case.

```
as.list(corpus)[1]
```

At this point you have a cleaned corpus of documents on which to base many other analyses. The entire script to this point will serve as the foundation of many other subsequent analyses performed in this and other chapters. Therefore it is important to master the reasons for which you bring in text, define it as a corpus and perform preprocessing steps.

2.8 Spellcheck

An optional preprocessing step that may be needed is correcting the spelling of terms. Text is often misspelled, and this next section will show one way to correct misspellings. You may want to correct the words depending on how impactful you expect misspelled words to be in your analysis. For example, legal documents and news articles likely do not contain a lot of misspelled words. In contrast, social media like Facebook generally contain a fair amount of misspelled terms. If you explicitly know expected misspellings such as "lol" for laugh out loud, then you may want to use the previously mentioned mgsub function. However, for more dynamic spell check you can use functions from qdap.

Consider Chapter 1's text mining definition with intentionally misspelled words.

```
tm.definition<-'Txt mining is the process of distilling
actionable insyghts from text.'
```

One way to check for misspelled words is with which_misspelled. The code below will identify the words not found in the basic qdap dictionary.

```
which_misspelled(tm.definition)
```

This returns a vector with the word position that is suspected of being misspelled and the word itself. The console output is shown here. Notice that "actionable" was thought to be misspelled so qdap's word dictionary is fairly limited. This function can help you identify words that are likely misspelled but does not help you correct them in the text.

```
1    7     8      9
"txt" "distilling" "actionable" "insyghts"
```

There is another function in qdap which allows you to interactively select a spelling correction from a list of possible terms. Appropriately named it is

```
check_spelling_interactive.
check_spelling_interactive(tm.definition)
```

One of the results of calling the interactive function on the text mining definition is shown below. The dialogue occurs in the console and requests a number corresponding to a correction choice. The choice dialogue repeats for every word that is not found in the qdap word dictionary. First, it shows the line containing a suspected misspelling. In this example, the specific word "actionable" is bracketed and thought to be incorrect. The next section in the console lists the available options and corresponding number. The last line is where you, as the user, type a number and hit enter to move to the next. Since "actionable" is in fact a word you would choose 2 and hit enter to move to the next word. In this example, the dialogue moves to <<distilling>>. In this case, you can select the appropriate correct spelling.

```
LINE: txt mining is the process of distilling
<<actionable>> insyghts from text.
SELECT REPLACEMENT:
  1: TYPE MY OWN         2: IGNORE: actionable    3:
atonable
```

```
    4: actiniae              5: accountable              6:
alienable
    7: auctioned             8: abominable               9:
affectionate
10: agitable            11: amicable
Selection:
```

This word by word interactive function can be useful in relatively small corpora. However, it is cumbersome when you have potentially tens of thousands of misspelled words as is the case with large Twitter corpora. As a result, a custom function can speed the process considerably.

```
fix.text <- function(myStr) {
  check <- check_spelling(myStr)
  splitted <- strsplit(myStr, split=' ')
  for (i in 1:length(check$row)) {
    splitted[[check$row[i]]][as.numeric(check$word.
no[i])] = check$suggestion[i]
  }
  df <- unlist(lapply(splitted, function(x) paste(x,
collapse = ' ')))
  return(df)
}
```

The `fix.text` function accepts a string, checks its spelling then replaces any suspect word with the first replacement from the suggestions from qdap's function. However, this is a classic speed versus accuracy tradeoff, so care must be taken in balancing your need to spellcheck accurately and the time which it would take to interactively spellcheck. This tradeoff is illustrated in the text mining example, because all words are corrected with the exception of "actionable," which is replaced with "atonable". Both the custom function application and console results are below. You must decide if changing words interactively is worth the accuracy improvement compared to the `fix.text` function.

```
fix.text(tm.definition)
[1] "text mining is the process of distilling atonable
insights from text."
```

The custom function can be applied to a single string or to an entire vector of documents. Given the limitations, it is best to avoid automatically performing spelling corrections, but it is nonetheless another tool in a text miner's toolset.

2.9   Frequent Terms and Associations

As previously shown, you can find out the existence of words within a corpus using the grep, grepl or stri_count. In those examples, the text was not transformed or cleaned. Often, cleaning the corpus will help aggregate terms so that a more accurate frequency count can be done. Also, simple frequency counts can often add insight yet do not require exotic mathematical techniques. Thus, performing a frequency analysis is often a good place to start when presented with a text mining project.

If you have followed along thus far, you should be able to create a clean corpus using the script in the previous section and the clean.corpus custom function. Further, recall that Table 2.2 is a term document matrix (TDM) to be used in the bag of words text mining methodology. When using the bag of words method, the matrix is what the analytics are based on. The next lines of code assume that you have a cleaned corpus from the previous section and then will convert your matrix into the TDM for analysis. First, you create a new object called tdm, which is a list object used by the tm package. There are different weighting schemes that can be used to create a TDM. Here you specify weightTf, which weights a TDM by term frequency. This is the default TermDocumentMatrix parameter which simply counts the number of occurrences by word. The other weighting options are document term inverse frequency, binary weights and weightSMART. These will be explored later in this book as we perform other analyses. After creating the TDM, you need to reclassify it as a matrix for easier analysis. The converted object name is tdm.tweets.m.

```
tdm<-TermDocumentMatrix(corpus,control=list(weighting
=weightTf))
tdm.tweets.m<-as.matrix(tdm)
```

First check the dimensions of the data frame. You will notice that it is 1377 columns and 2631 rows. This means that in total there are 2631 distinct words after performing the cleaning steps.

```
dim(tdm.tweets.m)
```

Due to the size of the matrix you may not want to use the head or tail functions to explore the matrix. Instead, you can index a specific couple of rows and columns to understand what this matrix contains. The code below returns rows 2250 to 2255 along with documents 1340 to 1342.

```
tdm.tweets.m[2250:2255,1340:1342]
```

Recall from an earlier keyword search that tweet 1340 contained the word "sorry." R will print the section of the matrix indexed in the line of code above and show you the exact tweet that mentions "sorry." It is marked with a one because that tweet contained one instance of that term. Figure 2.3 shows the section of the matrix with a tweet containing "sorry."

Figure 2.3 The section of the term document matrix from the code above.

	Docs		
Terms	1340	1341	1342
sooo	0	0	0
sootawn	0	0	0
sophiesoph	0	0	0
soraparuq	0	0	0
sorry	0	0	0
sound	0	0	0

At this point, you may also recognize that the matrix contains a lot of zeros. This is often the case in this text mining method. From a practical point of view it should not be a surprise. Here, a tweet may average 10 distinct words. As a result each column may only have ten or so numbers. Consider that there are more than 2000 different terms and you are left with a bunch of zeros. Although all of these tweets are from DeltaAssist, the reality is that language choice is diverse, and so both TDM and DTM are often filled with many zeros, making them sparse.

In order for you to summarize the frequency of terms, you will need to sum across each row because each row is a unique term in the corpus. You need to call the base R function rowSums and pass the matrix to it.

```
term.freq<-rowSums(tdm.tweets.m)
```

Then you create a new data frame object by putting the original terms and the term frequencies next to each other. The new freq.df object has two columns, one for terms and another that is just the summation for each row.

```
freq.df<-data.frame(word=names(term.freq),
frequency=term.freq)
```

Now you can sort the data frame and then look at the first ten most frequent terms. If you do this, you will see a lot of DeltaAssist tweets have please, sorry, flight confirmations, and numbers. As an airline, this may not be surprising, but it can still be insightful to understand common issues and help to draw inferences.

```
freq.df<-freq.df[order(freq.df[,2], decreasing=T),]
freq.df[1:10,]
```

2.10   DeltaAssist Wrap Up

Given the text mining skills you have learned thus far, you should be able to answer some of the original business case questions I had when starting the social media customer service group at Amazon. At the time, we were very concerned with legal, marketing and brand, and publicly made promises, so we decided to understand the types of replies some of the companies we admired in the space were writing. Using simple functions like `nchar`, `grep`, `grepl`, and string counts, we could identify typical length of tweets, and look for consistently used patterns like urls or customer service agent signatures. Reviewing the most frequent terms can provide some insight into typical customer services issues these companies encounter. That helped us prepare for and think about our own most likely issues. In this chapter, I only discussed single word tokens, but later you will investigate n-gram tokenization, specifically looking at bi-grams or two-word combinations. Doing so will enrich the insights that can be extracted, but the underlying bag of words methodology remains the same.

2.11   Summary

In this chapter you learned:

- what text mining is
- basic low level text mining functions
- a business use case that benefited from basic text mining
- the bag of words R script that is foundational to many other beginning text mining analysis
- frequency analysis
- summarizing word frequency

3

Common Text Mining Visualizations

In this chapter, you'll learn

- to visualize a simple bar plot of word frequencies
- to find associated words and make a related plot
- to make a basic dendrogram
- to make and improve the aesthetics of a hierarchical dendrogram showing basic word clustering
- how to make a word cloud
- how to compare word frequencies in two corpora and create a comparison cloud
- how to find common words and represent them in a commonality cloud
- how to create a polarized cloud to understand how shared words "gravitate" to one corpus or another
- how to construct a word network quickly

3.1   A Tale of Two (or Three) Cultures

So far in this book you have learned about some very basic text mining that was done when launching a social media team at Amazon. At its heart Amazon is a book company, and senior stakeholders consume information through narrative. For three years, each quarter I had to create a quarterly business review, QBR, which was a six-page paper. This document was almost entirely narrative form explanations of the past quarter's business issues and successes. In it, charts and tables were relegated to the appendix and seldom looked at. It is an Amazon management belief that in order to understand your section of the business at a deep level and not be misleading or full of "fluff," you need to write out business topics. While they do not stand on agendas, large audiences or memos, the fact remains that Amazon's stakeholders are committed to concise narratives in these QBRs. During many presentations, my bosses and theirs

Text Mining in Practice with R, First Edition. Ted Kwartler.
© 2017 John Wiley & Sons Ltd. Published 2017 by John Wiley & Sons Ltd.

would sit reading my work quietly while I passed the time waiting for questions. There was no sense in rereading my own work just to appear busy! After about 20 minutes of silence, the questions would come and they would come fast. After an early failed QBR, in which I struggled to answer the probing questions, I understood the importance of what the QBR narratives were meant to illustrate to leadership.

After Amazon, I joined a different Fortune 100 company that had a significantly different culture. In fact, there was little text for the senior stakeholders to read during meetings. In contrast, it was a visualization-heavy culture. Most, if not all, of the meetings had PowerPoint visuals to convey meaning, allowing the directors, vice-presidents and so on to understand the business issue and then talk it through in a meeting. The collaboration and respect given to all in attendance was impressive, and a good solution usually followed. As was the case at Amazon, I originally misunderstood my audience's needs. My first presentations had text-heavy slides, and I had sent lengthy emails to participants for review prior to the meetings. To my surprise no one had read the emails, and as the text-heavy slides wore on, the participants abandoned the slides altogether in favor or talking through the topic collegially. Over the course of the next two years I worked to hone my craft for this new visual-heavy culture. It was the only way to gain alignment on core issues with senior decision-makers.

The two cultures could not have been more different, yet both were effective in their domain. Despite being a fast-moving technology company, Amazon surprisingly relies on text. On the other hand, the decades-old seemingly less agile yet equally successful company in a different industry favored visualizations and talking through topics among decision-makers.

After that I joined a small, venture-backed startup. The organization was full of kaggle.com's top-ranked data scientists. This data science heavy organization again had a different culture. As an extremely data fluent organization building an extremely sophisticated machine learning platform, the audience did not rely on long form narratives to execute a plan or understand an issue. Also, visuals were kept to a minimum in favor of tabled and organized data. The creative minds of this data science culture, with PhDs from top institutions worldwide had no trouble comprehending lengthy data in tables. To them the story was in the numbers, and if the numbers were not correct, the rest was incorrect, no matter the explanation.

Depending on the work culture you find yourself in, you may favor insights based on the text mining as, explanations with tabled numerical data of the previous chapter or visuals created in this chapter. Visualizations can be a powerful method to convey meaning. So if you find yourself in a culture that relies on images to process information, then this chapter is a good starting point for your text mining efforts. The book has other visualizations, but the ones in this chapter are foundational to many text-mining efforts.

> **Getting the Data**
>
> This chapter assumes that you were able to follow along for the bag of words organization articulated in the previous chapter. That means you should be able to load a simple CSV file, clean and preprocess it and then create either a term document matrix or document term matrix. If you struggled with these concepts, then review them once more before proceeding. This chapter rapidly moves from file to document matrix, so that explanations are kept to the additional manipulations needed to create the visualizations. In order to create the exact visualizations in this chapter, go to www.tedkwartler.com and download oct_delta.csv and amzn_cs.csv. Throughout this chapter you will use both files.

3.2 Simple Exploration: Term Frequency, Associations and Word Networks

Sometimes merely looking at frequent terms can be an interesting and insightful endeavor. On some occasions frequent terms are expected within a text mining project. However, unusual words or (later in the book as you explore multi-gram tokenization) phrases can yield a previously unknown relationship. This section of the book constructs simple visualizations from term frequency as a means to reinforce what would probably already be known in the case of DeltaAssist's customer service tweets. It then goes a step further to look at a specific term association. Text mining's association is similar to the statistical concept of correlation. That is to say, as the frequency of a single word occurs, how correlated or associated is it with another word? The exploration of the term association can yield interesting relationships among a large set of terms. Without also coupling association with word frequency, this may actually be misleading and become a fishing expedition, because the number of individual terms can be immense. Lastly this section adds a word network using the qdap package. This is another way in which to explore association and connected terms. Those familiar with social network analysis will be equally familiar with the concept of a word network. This relationship between words is captured in a special matrix called an adjacency matrix, similar to the individuals of a social network. A word network differs from word association. A word network explores multiple word linkages simultaneously. For example, the words "text," "mining" and "book" can all be graphed at the same time in a word network. The word network will have scores for pairs "text" to "mining," "text" to "book" and "mining" to "book". In contrast, word association scores represent the relationships of a single word to others, such as "text" to "mining" and "text" to "book". This contrasts because there is no score for the pair "mining" and "book."

3.2.1 Term Frequency

Although not aesthetically interesting, a bar plot can convey amounts in a quick and easy-to-digest manner. So let's create a bar plot of the most frequent terms, and see if anything surprising shows up. To do so you will be loading the package ggthemes. This package has predefined themes and color palettes for ggplot2 visualizations. As a result, we do not have to specify them all explicitly. Using it saves time compared to using the popular ggplot2 package alone. There are other visualization packages within the R ecosystem but ggplot2 is both popular and adequate in most cases.

In review from the previous chapter you need to bring in a corpus and then organize it. To do so you ultimately need to get back to a cleaned term document matrix. After applying last chapter's clean.corpus custom function, you need to make the matrix and then, as before, get the row sums into an ordered data frame. The code below should look very familiar as it redoes the same steps as the previous chapter, ending in an ordered data frame of term frequencies. However, at this point we are going a step beyond the tabled data and create a simple bar plot.

```
library(tm)
library(ggplot2)
library(ggthemes)
text.df<-read.csv('oct_delta.csv')
tweets<-data.frame(ID=seq(1:nrow(text.df)),
text=text.df$text)
tryTolower <- function(x){
  y = NA
  try_error = tryCatch(tolower(x), error = function(e) e)
  if (!inherits(try_error, 'error'))
    y = tolower(x)
  return(y)
}


custom.stopwords <- c(stopwords('english'), 'lol',
'smh', 'delta', 'amp')
clean.corpus<-function(corpus){
corpus <- tm_map(corpus,
content_transformer(tryTolower))
corpus <- tm_map(corpus, removeWords,
                custom.stopwords)
corpus <- tm_map(corpus, removePunctuation)
corpus <- tm_map(corpus, stripWhitespace)
```

```
corpus <- tm_map(corpus, removeNumbers)
return(corpus)
}
meta.data.reader <- readTabular(mapping=
list(content="text", id="ID"))
corpus <- VCorpus(DataframeSource(tweets),
readerControl=list(reader=meta.data.reader))
corpus<-clean.corpus(corpus)
tdm<-TermDocumentMatrix(corpus,control=list(weighting
=weightTf))
tdm.tweets.m<-as.matrix(tdm)
term.freq<-rowSums(tdm.tweets.m)
freq.df<-data.frame(word=
names(term.freq),frequency=term.freq)
freq.df<-freq.df[order(freq.df[,2], decreasing=T),]
```

All of the above code is needed to create the `freq.df` data frame object. This becomes the data used in the `ggplot2` code below that constructs the bar plot. In order to have `ggplot2` sort the bars by value, the unique words have to be changed from a string to a factor with unique levels. Then you actually call on `ggplot2` to make your bar plot. The first input to the `ggplot` function is the data to reference. Here you specify only the first 20 words so that the visual will not be too cluttered. The `freq.df[1:20,]` below can be adjusted to add more or fewer bars in the visualization.

```
freq.df$word<-factor(freq.df$word,
levels=unique(as.character(freq.df$word)))
ggplot(freq.df[1:20,], aes(x=word,
y=frequency))+geom_bar(stat="identity",
fill='darkred')+coord_flip()+theme_gdocs()+
geom_text(aes(label=frequency),
colour="white",hjust=1.25, size=5.0)
```

The gg within `ggplot` stands for grammar of graphics and creates the visualization. The `ggplot` code uses a structured and layered method to create visualizations. First we pass the data object to be used in the visual `freq.df`, indexing only the first 20 terms. Next we define the aesthetics with the x and y axis referencing the column names of the data object. Once this is done we add a layer using the plus sign. The new layer is to contain bars, and so it uses the `geom_bar` function. It creates one-dimensional rectangles whose height is mapped to the value referenced. Further, `geom_bar` must be told how to statistically handle the values. Here you specify the identity so that each bar represents a unique identity and is not transformed in another manner. The code also fills the bars with dark red. You can specify various colors, including hexadecimal colors, as part of the fill parameter. The next layer is again added with a plus sign and simply rotates the x and y of the graph. This is an empty function call and can be removed if it suits the analyst making the visual. The next layer is actually from `ggthemes` and represents an entire predefined style. In this case, it is meant to mimic Google document visualizations. You can change the style manually using many parameters, leave the default `ggplot2` style or use a `ggtheme` as you have done here. Lastly, another layer is added on top of the bars. The `geom_text` layer represents the white numerical text labels at the end of each bar. On this last layer, you can change color, adjust the position, and even adjust it to the size you desire. Figure 3.1 shows the results of these efforts to construct a bar plot of frequent terms.

In this rudimentary view, you can see that many of the tweets from Delta are apologies and discussions about flight and confirmation numbers. There is nothing overly surprising or insightful in this view, but sometimes this simple approach can yield surprising insights. Later you will use two word pairs instead of single word tokens. In my experience, changing the tokenization can enrich the insights that are found through a basic term frequency visual. The website has an extra Twitter data set called chardonnay.csv in which this approach can show an unexpected yet frequent result as you adjust the stopwords.

Notice that all the words are lowercase, and there is even a mistakenly parsed word "amp." This is the result of the character encoding not being recognized

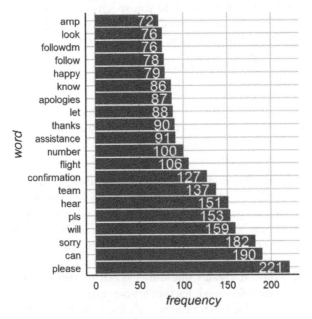

Figure 3.1 The bar plot of individual words has expected words like please, sorry and flight confirmation.

properly. Character encoding is the process of converting text to bytes that represent characters. R needs to read each character and encode it to a corresponding byte. There are numerous encoding types for languages, characters and symbols, and as a result mistakes can occur. The same issue can occur with emoticons, as those are parsed into completely different characters and byte strings than would be necessary for R to be able to make sense of them. In subsequent scripts you will add more lines of code to specify encoding, thereby changing the "amp" to the ampersand "&."

Based on this initial visualization, you can dive further into the analysis. In the bar chart, apologies is mentioned many times. Sometimes it makes sense to explore unexpected individual words and find the words most associated with them. In this simple example, you can explore associated terms with apologies to hopefully understand what DeltaAssist is apologizing for. In your own text mining analysis, the surprising words in the bar plot or frequency analysis are ripe for the following additional exploration called "association."

3.2.2 Word Associations

In text mining, association is similar to correlation. That is, when term x appears, the other term y is associated with it. It is not directly related to frequency but instead refers to the term pairings. Unlike statistical correlation, the range is between zero and one, instead of negative one to one. For example, in this book the term "text" would have a high association with "mining," as we refer to text mining often together.

The next code explores the word associations with the term "apologies" within the DeltaAssist corpus. The word "apologies" was chosen after first reviewing the frequent terms for unexpected items, or in this case, to learn about a behavior of customer service agents. Since the association analysis is limited to specific interesting words from the frequency analysis, you are hopefully not fishing for associations that would yield a non-insightful outcome. Since all words would have some associative word, looking at outliers may not be appropriate, and thus the frequency analysis is usually performed first. In the next code, we are looking for highly associated words greater than 0.11, but you will likely have to adjust the threshold in your own analysis.

The code itself creates a data frame of factors for each term and the corresponding association value. The new data frame has the same information as the association's object matrix but is a data frame class for `ggplot2`. This data frame has a superfluous vector with rows names as a vector. The data frame also changes the terms from strings to categorical factors. These steps may seem redundant, but this approach makes it explicit and easy to follow when the data is used in `ggplot`.

```
associations<-findAssocs(tdm, 'apologies', 0.11)
associations<-as.data.frame(associations)
associations$terms<-row.names(associations)
associations$terms<-factor(associations$terms,
levels=associations$terms)
```

Once you have a clean data frame of highly associated words and their corresponding values, you can use it for another visual. Again building by layer ggplot is pulling data from the "association" data frame. You are setting the y axis to be the terms and the x axis to be the values. Instead of geom_bar, you are using geom_point and setting the size explicitly. Next you use the predefined gdocs theme. Then you add a layer of labels for the values in dark red. Lastly you change the default gdoc theme by increasing the y axis label's size and removing the y axis title. The code below creates Figure 3.2, showing the most associated words with "apologies" in our corpus.

```
ggplot(associations, aes(y=terms)) +
geom_point(aes(x=apologies), data=associations,
size=5)+
   theme_gdocs()+ geom_text(aes(x=apologies,
label=apologies),
colour="darkred",hjust=-.25,size=8)+
   theme(text=element_text(size=20),
axis.title.y=element_blank())
```

Again, notice some poor parsing of the text done by R. Instead of "you're", R has interpreted the word to include some foreign characters and even a trademark

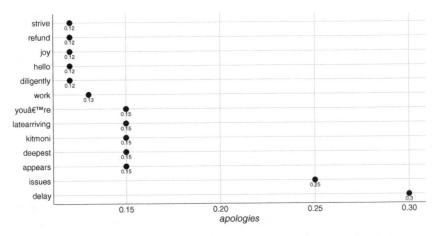

Figure 3.2 Showing that the most associated word from DeltaAssist's use of apologies is "delay".

abbreviation! You will finally learn how to clean this up in the word cloud section, but for now focus on the meaning of association and basic visualization.

In this case, these words confirm what you likely already know, that airline customer service personnel have to apologize for late arrivals and delays. However, in other instances this type of analysis can be useful. Consider a corpus with technical customer reviews and complaints for laptops. Performing a simple word frequency and association analysis may yield the exact cause of poor reviews. You could find common words – e.g. "screen problem" – within the corpus. And reviewing the associated words with screen and problem may yield highly associated terms like hdmi and cable or driver. It is often the case that term frequency and word association alone can yield some surprising results that can lead to an insight or confirm an existing belief. In this simplistic case the following tweet confirms the word frequency and association conclusion that agents apologize for delays.

"@kitmoni At the moment there appears to be a delay of slightly over an hour. My apologies for today's experience with us. *RB"

3.2.3   Word Networks

Another way to view word connections is to treat them as a network or graph structure. Network structures are interesting in conveying multiple types of information visually. Word networks are often used to demonstrate key actors or influencers in the case of social media. Within the context of text mining, networks can show relationship strength or term cohesion, leading to an assumption of a topic. A word of caution, as these can become dense and hard to interpret visually. As a result, it is important to restrict the number of terms that are being connected. In the earlier analysis, you saw that the words "apologies" and "refund" are highly associated. A word network may more broadly indicate under what circumstances Delta would issue a refund. Since the term document matrix contains thousands of words, in your example you limit the network illustration to the word "refund." Word networks can be used to understand word choice by visually producing clusters in the layout. Further, sometimes entire topics can be interpreted visually based on these diagrams. In a later chapter we cover other clustering techniques but this is a qualitative, audience-based approach that is worth learning.

A simple network map is shown in Figure 3.3. The lines connecting the circles in a network graph are called edges. The circles themselves are called nodes or vertices. A network graph can have many dimensions of information contained in it. The example below has the same size nodes, and edge thickness. However, some of the parameters that can be adjusted in word networks include the size of the nodes often showing more prominent members of the network, thickness of lines representing the strength of a connection, and of course color, which can denote particular class attribution such as a race or

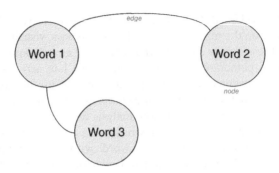

Figure 3.3 A simple word network, illustrating the node and edge attributes.

gender. Since there are no further informational dimensions applied, Figure 3.3 merely shows that word 1 is connected with both words 2 and 3 but that the nodes representing words 2 and 3 are not.

To create a word network graph, an R user can employ the igraph library, which strives to provide "pain-free implementation" of graph structures. In order to build a word network, you first need to limit the term document matrix; otherwise the network will be too dense to meaningfully compre-hend. So, the code below uses grep and a specific pattern match to index the entire tweets data. In this case, the pattern, refund, was chosen based on the frequent terms and association analyses performed earlier. Using grep to index the original tweets data frame leaves only seven tweets. To further reduce clutter, the code below indexes only the first three of the refund-men-tioning tweets to build the corpus. This may be too few to have credibility in a practical application but nonetheless it is a good example of how to build a simple word network graph. Later, for hierarchical clusters, the function to remove sparse or infrequent terms is introduced. Removing sparse terms mathematically is another way of decreasing the size of a TDM. The object refund is created using grep and represents seven tweets from the original tweets data.

```
library(igraph)
refund<-tweets[grep("refund", tweets$text,
ignore.case=T), ]
```

As before, you move from a data frame to the bag of words style matrix using functions from the tm package, and refund.reader reads the table of refund data and then turns it into a volatile corpus. Next the small corpus is cleaned using the prior custom function clean.corpus. Lastly, a refund.tdm object is made by calling the term document matrix function.

```
refund.reader <- readTabular(mapping=list(content="text",
id="ID"))
refund.corpus <- VCorpus(DataframeSource(refund[1:3,]),
            readerControl=list(reader=refund.reader))
refund.corpus<-clean.corpus(refund.corpus)
refund.tdm<-TermDocumentMatrix(refund.corpus,
                control=list(weighting=weightTf))
```

Next we need to create an adjacency matrix, which is a simple matrix with the same row and column names, making it square. At the intersections, there is a binary operator, 1 or 0, showing a connection or not. While this may seem confusing, consider the following simple example starting with a term document matrix. Table 3.1 is a small term document matrix of fictitious terms called `all`.

Notice at this point that the matrix is not square. There are more tweets than there are words. In reality, the opposite may be true, but it does not matter because the next step transitions from this table to an adjacency matrix which is square. In order to create an adjacency matrix, R's binary operators need to be used on the term document matrix. Specifically the matrix multiplication binary operator is called. To the novice R programmer, binary operators may seem tricky, but they are in fact fairly straightforward. The pseudo code for a binary operator is `first input %some function% second input`. The code below illustrates just the matrix multiplication operator from among the other available ones, such as `modulo` and `match`. The first input is 10, followed by the binary operator and then ending with a 2.

```
10%*%2
```

This will return a matrix with one row and one column. In the first and only cell is the answer 20, as shown in Figure 3.4.

Binary operators are useful in a broad application. The next example demonstrates the matrix multiplication binary operator applied to more than one number at a time. The result is a matrix with two columns with answers 10

Table 3.1 A small term document matrix, called `all` to build an example word network.

	Tweet1	Tweet2	Tweet3	Tweet4	Tweet5	Tweet6
R	0	0	1	1	0	0
Text	0	0	1	0	0	0
Stats	1	1	0	0	0	0
Mining	0	1	1	0	0	0
Book	0	0	0	0	1	1

```
> 10%*%2
     [,1]
[1,]   20
```

Figure 3.4 The matrix result from an R console of the matrix multiplication operator.

times 2 and 10 times 3 in each cell. R's console output to this operation is captured in Figure 3.5.

```
10%*%c(2,3)
```

```
> 10%*%c(2, 3)
     [,1] [,2]
[1,]   20   30
```

Figure 3.5 A larger matrix is returned with the answers to each of the multiplication inputs.

Moving back to our example of the small term document matrix in Table 3.1, you can apply the same operator on the original TDM and the transposition of the TDM. This will make the matrix square in a new object called 'adj.m'.

```
adj.m <-all %*% t(all)
```

Reviewing the original TDM, `all`, note that "R" is shared in tweets 3 and 4. Tweet 3 contains the words R, Text and Mining. So we should expect that R will be connected to both Text and Mining, and when comparing R to itself, there is a loop or redundant connection. Similarly, tweet 4 has only the term R and does not share any other words in the TDM. As a result, tweet 4 will have no other external connections. However, when comparing tweet 4 to itself, it shares the term R and again there is a redundant loop. More explicitly, the intersection of R and R has a 2, representing the loop for each of the tweets 3 and 4. The intersection of row R and column Text contains a 1 because the original data frame has a 1 at row Text and the Tweet 3 column, and there is also a 1 at row R in the same Tweet 3 column. Further, reviewing the `all` TDM more closely, all terms are in at least two tweets with the exception of the term Text. When looking at

Table 3.2 The adjacency matrix based on the small TDM in Table 3.1.

	R	Text	Stats	Mining	Book
R	2	1	0	1	0
Text	1	1	0	1	0
Stats	0	0	2	1	0
Mining	1	1	1	2	0
Book	0	0	0	0	2

the diagonal values in the resulting `adj.m` object in Table 3.2, all terms have a 2 with the exception of the Text and Text intersection. These represent the redundant loops. Overall, the matrix multiplication operator is applied to each of the terms and corresponding tweets to get the complete result in Table 3.2.

You are safe to build the refund adjacency matrix using the matrix multiplication operator, now that the fundamental construction has been explained. The `refund.tdm` object from earlier in this section needs to be converted to the appropriate matrix class. The matrix multiplication operator is applied as before to this new matrix object. Then the object is transformed into an object that is used by the `igraph` library. The last line removes the loops, as they are more visual clutter and do not provide additional context to the DeltaAssist refunds.

```
library(igraph)
refund.m<-as.matrix(refund.tdm)
refund.adj<-refund.m %*% t(refund.m)
refund.adj<-graph.adjacency(refund.adj, weighted=TRUE,
mode="undirected", diag=T)
refund.adj<-simplify(refund.adj)
```

Plotting in the `igraph` package can be challenging because there are many input parameters such as shape, thickness and color for both edges and vertices. That said, a simple word network graph that can be built using the refund adjacency matrix is shown in Figure 3.6. The refund adjacency matrix is passed to the `plot.igraph` function. The following parameters help to alleviate the visual clutter. Vertex shapes, such as circles, are removed. Then the label font is passed, so that the font itself is not overtly busy. The label colors are red, and the size is adjusted to be somewhat smaller than the default. Lastly the edge lines are colored with a faint gray. By doing so, the terms are more easily read but edge connections are not completely lost. Reversing the color scheme for edges and nodes would mean that the edges would become the focal point, and the terms may be hard to decipher. So care must be taken when building this type of visual, so that the audience can draw meaning. Figure 3.6 comprises an adjacency matrix of three tweets.

```
plot.igraph(refund.adj, vertex.shape="none",
     vertex.label.font=2, vertex.label.color="darkred",
     vertex.label.cex=.7, edge.color="gray85")
title(main='@DeltaAssist Refund Word Network')
```

The result of the simple word network shows a strong connection between refund and apologies. Since this is based on only three tweets, it should not be a surprise that there are three distinct network clusters. The first two are linked

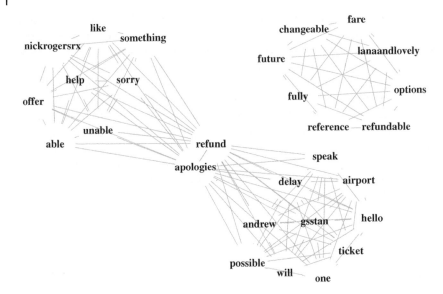

Figure 3.6 A very small word network using the `igraph` package.

by the words "apologies" and "refund." This appears to confirm the associative relationship between the words as seen previously. Still the third tweet stands alone. This is because it has the word refundable which was captured using the original `grep` indexing, is technically a different term than "refund," so no network connection was created linking all three. In the word cloud section of this chapter, we cover word stemming and spellcheck, which both help further aggregate terms to avoid this issue.

The qdap package provides a convenient wrapper to create this type of visual. The package author, Tyler Rinker, has also selected some attractive and common-sense plot parameters, making the use of the functions very stress free. In fact, the functions explained next do not require you to manually create the term document matrix! Going through the manual exercise of creating an adjacency matrix helps to ensure comprehension, but using qdap's `word_network_plot` and `word_associate` functions saves considerable time and effort. The single line of code to create Figure 3.7 below essentially creates the exact same visual as Figure 3.3!

```
library(qdap)
word_network_plot(refund$text[1:3])
title(main='@DeltaAssist Refund Word Network')
```

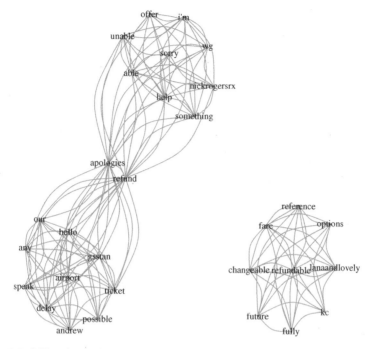

Figure 3.7 Calling the single qdap function yields a similar result, saving coding time.

qdap's word association network function goes another step further by utilizing another binary operator, the `%in%` or `match` function. The function is passed the entire corpus of tweets and then a pattern upon which to find matches. In this example, when the `refund` pattern is found, the matching operator underneath returns a `TRUE`. As the function progresses, all pattern matches are kept for building the network visual while all others tweets are discarded. In doing so, the function mimics the `grep` indexing performed earlier, saving another line of code. All of the underlying data structures to create the adjacency matrix and ultimately the network visual are contained by calling this single function. In Figure 3.7's code, the goal is to match tweets containing "refund" from among the entire corpus. In the code, the match.string parameter "refund" is within a concatenate function. If needed, additional string patterns can be input by adding a comma and another quoted pattern between the parentheses. In the previous example, the code limited the refund data frame from seven tweets to only the first three. Calling this function will use all seven tweets that return a `TRUE` for the pattern match. As a result, Figure 3.8 will be slightly different and more cluttered than the others because it is based on more information.

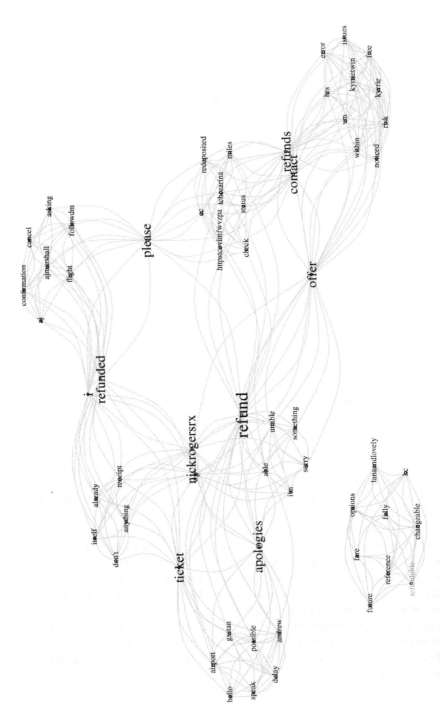

Figure 3.8 The word association function's network.

```
word_associate(tweets$text,match.string=c('refund'),
        stopwords=Top200Words,network.plot = T,
        cloud.colors=c('gray85','darkred'))
title(main='@DeltaAssist Refund Word Network')
```

Tip: Notice that in both qdap applications the basic cleaning steps from the custom `clean.corpus` function were applied without explicitly calling them. This can be a blessing of saving time but can also lead to less explicit control.

In conclusion, word networks used on small corpora or in conjunction with other basic exploratory analyses can be helpful. In this case, the takeaways are minimal because the amount of information that the word networks are based upon was purposefully limited. Further, as your corpus grows and term diversity increases, word networks will likely become less and less impactful. Nonetheless, word networks represent a basic text mining visualization that is worth adding to an explorative text project but possibly omitted for final presentations given their frenetic nature.

3.3   Simple Word Clusters: Hierarchical Dendrograms

Hierarchical dendrograms are a relatively easy approach for word clustering. Later, you learn more complex clustering, but this section will provide a basic means of extracting meaningful clusters. A dendrogram is a tree-like visualization and in this case based simply on frequency distance. This analysis is an information reduction. Another example of an information reduction is taking the average for a population. We reduce information to reach an amount of information that we can more readily understand about the whole population. Consider the following table of rainfall data from www.weatherdb.com. Table 3.3 is a small data set of city rainfall upon which we can then build Figure 3.9's dendrogram.

The dendrogram, Figure 3.9, reduces the information of the rainfall data to explain similarities by city. Cleveland and Portland are at the same height

Table 3.3 A small data set of annual city rainfall that will be used to create a dendrogram.

City	Annual rainfall (in)
Boston	43.77
Cleveland	39.14
Portland	39.14
New Orleans	62.45

Annual Rainfall Dendogram

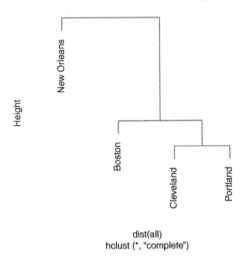

Figure 3.9 The city rainfall data expressed as a dendrogram.

because their distance measures are 0. Boston receives a bit more rain, so it is set apart and above, yet is closer in difference than New Orleans. New Orleans receives so much rainfall that it is the highest city and also set well apart from the other cluster.

A similar visualization can be created using a TDM, so you can visually explore token frequency relationships instead of city rainfall. However, unsophisticated frequency based similarity plots often help identify phrases or topics that warrant further exploration. Previously, you created a TDM to explore simple word frequency and association expressed as visuals. To create a dendrogram, you diverge from those visualizations after making the TDM. Recall from the previous section that the matrix version of the TDM using the DeltaAssist corpus had 2631 terms and 1377 rows or tweets. You need to reduce this sparse matrix considerably in an effort to make the visualization comprehensible. In my experience, it is best to reduce the TDM to approximately 50 distinct terms to make a worthwhile dendrogram. Given the lexical diversity of most corpora, usually some clusters are not helpful but there are some that can provide insights.

Previously we used `grep` to index tweets and then a string pattern parameter within a function. Instead of reducing a matrix size based on a string pattern, you can reduce the dimensions mathematically. To reduce the tdm we apply the `removeSparseTerms` from the `tm` package. The manner in which this function works is by supplying first a TDM or DTM and then a sparse parameter.

The sparse parameter is a number between 0 and 1. It measures the percentage of zeros contained in each term and acts as a cut-off threshold. For instance, supplying a sparse parameter of 0.99 would include all terms with 99% or fewer zeros. In contrast, changing the parameter to 0.10 would indicate that only terms that have 10% or fewer zeros are retained among all documents. In reality, many corpora are likely to have 0.95 or more zeros, so it is good to tune your dendrogram starting at 0.95 to 0.999999. To create your new TDM object called tdm2, follow the next code. You can compare the original tdm to the tdm2 by typing tdm then tdm2 into your console to see the summarized results and how many terms have been reduced. If you do so, you will notice that the new tdm object, tdm2, has only 43 terms.

```
tdm2 <- removeSparseTerms(tdm, sparse=0.975)
```

Once satisfied with the tdm2 size being between 40 and 70 terms, you should be able to perform a hierarchical cluster analysis by measuring the distance between term vectors. The dist function creates a difference matrix by computing a distance between the vectors. The default way to measure distance is Euclidean, as shown, but you may specify other measures including maximum or binary. It may be worthwhile to change to these other distances to see how the distance measures affect the shape of the tree visual. The distance matrix is then passed to the hclust function to collect the needed information in a list. The hclust function first assumes each term is its own cluster and then iteratively attempts to join the two most similar clusters. Ultimately all individual clusters, then grouped clusters, are merged into one single cluster, and distance measures are calculated again. As with the dist function, there are a number of different clustering methods. The default is the complete method, but you may specify another, such as centroid or median. Some text mining practitioners prefer to use the median or mediod clustering techniques because the lexical diversity leads to outliers that can affect a mean clustering approach.

```
hc <- hclust(dist(tdm2, method="euclidean"),
method="complete")
```

The last line of code to create your first dendrogram clustering visualization is to merely plot it. We call the plot function, then pass the hc object made in the line before. Lastly, we remove the y axis and add a title. I prefer to remove the y axis as the height numbers appear very low to stakeholders and they may focus on that more so than on the informative clusters. The height is representative of the distance measures at this point (not the term frequencies), so it may be misleading. Often just adding these three lines of code after cleaning a corpus and creating a TDM can create an insightful cluster analysis. Figure 3.10 represents a plot of the hc object with the "@DeltaAssist Dendrogram" title.

```
plot(hc,yaxt='n', main='@DeltaAssist Dendrogram')
```

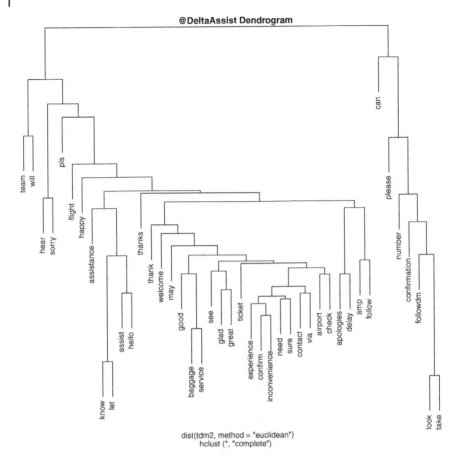

Figure 3.10 A reduced term DTM, expressed as a dendrogram for the @DeltaAssist corpus.

The dendrogram here shows a lot of apologetic wording and a distinct cluster asking for a direct message. There is also a smaller yet distinct cluster related to baggage service, so it looks like a fair number of tweets were not only related to delays but also to baggage service.

Although informative in some contexts, the base dendrogram may not be as visually pleasing as hoped. The following lines of code simply add color, slightly change the shape and align all words across the bottom. You can create a custom function called dend.change. First you pass in an object. If the object is a leaf or end point of the dendrogram, it grabs the attributes of that specific leaf. Then it assigns the color labels to the attributes of the leaf, based on the node or cluster.

```
dend.change <- function(n) {
  if (is.leaf(n)) {
```

```
  a <- attributes(n)
  labCol <- labelColors[clusMember[which(
    names(clusMember) == a$label)]]
  attr(n, "nodePar") <- c(a$nodePar, lab.col =
                          labCol)
 }
 n
}
```

Once you have the coloring function, you apply it in your code to improve the visual. First, you reclassify the hc object from a hierarchical cluster object to a dendrogram. Then you cut the original hierarchical cluster object into four distinct groups to be used for the plot. In more diverse corpora, you may want to increase the cutree parameter beyond four. When building this type of visualization it may make sense to try different clusters within the cutree function. Next, you need to specify the colors of the groups when you create the labelColors object. To specify colors, you can use hexadecimal or basic color names inside the parentheses. The total number of colors must be equal to the number of clusters specified in the previous function. Also, the colors are used in the order in which they are coded. In the example below, the first cluster is assigned a darkgrey color and so on. Next, we apply the custom coloring function to the hcd dendrogram object. The custom coloring function actually needs an object called clusMember, which we created earlier. This object denotes how terms are grouped when we dendrapply the custom dend. change function. The last step merely plots Figure 3.11 with the previous title and changes from a rectangular line shape to a triangle view.

```
hcd <- as.dendrogram(hc)
clusMember <- cutree(hc,4)
labelColors <- c('darkgrey', 'darkred', 'black',
'#bada55')
clusDendro <- dendrapply(hcd, dend.change)
plot(clusDendro, main = "@DeltaAssist Dendrogram",
type = "triangle",yaxt='n')
```

Still, this view may not be pleasing or visually convey the important clusters well. Instead, we can use libraries specifically tailored to improve the aesthetics of dendrograms. Specifically, Tal Galili's excellent package dendextend combined with the circlize package can create an unusual and attention-grabbing version of the dendrogram. In Figure 3.12, we change the color order so that we call out the specific clusters to convey the interesting insight about @DeltaAssist.

Adjust the hcd dendrogram object after loading both the dendextend and circlize libraries. Using dendextend functions cut the tree to four

@DeltaAssist Dendrogram

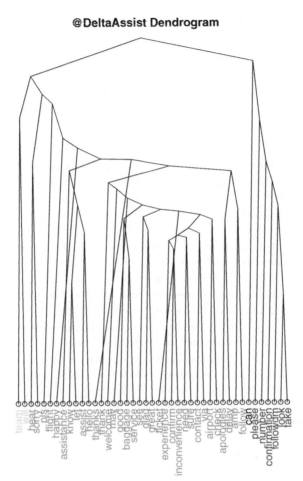

Figure 3.11 A modified dendrogram using a custom visualization. The dendrogram confirms the agent behavior asking for customers to follow and dm (direct message) the team with confirmation numbers.

clusters then adjust the color labels explicitly. Second, the code below mimics the number of clusters, and colors but is applied to the dendrogram's branches. This will help the consistency of the visual. Lastly, instead of calling the base plot function, call the `circlize_dendrogram` function to create a novel illustration in Figure 3.12.

```
library(dendextend)
library(circlize)
hcd<-color_labels(hcd,4, col = c('#bada55','darkgrey',
"black", 'darkred'))
```

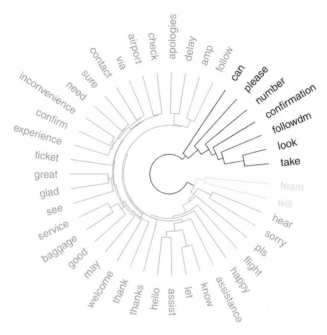

Figure 3.12 The circular dendrogram highlighting the agent behavioral insights.

```
hcd<-color_branches(hcd,4, col = c('#bada55','darkgrey',
"black", 'darkred'))
circlize_dendrogram(hcd, labels_track_height = 0.5,
dend_track_height = 0.4)
```

3.4 Word Clouds: Overused but Effective

Another common visualization is called a word cloud or tag cloud. Generally, a word cloud is a visualization based on frequency. In a word cloud, words are represented with varying font size. In a simple word cloud, only one dimension of information is shown. Specifically, the font size corresponds to n-gram frequency. That means that the larger a word in the word cloud the more frequent the word is in the corpus. Other dimensions of a word cloud can be changed to demonstrate new information, such as color and grouping. This section demonstrates the word cloud package and introduces a polarized tag cloud in a pyramid plot. The pyramid plot is constructed using the `plotrix` package.

In general, word clouds are popular because audiences can easily comprehend the illustration. This has led to an over use of word clouds during text mining projects. Also, you can manipulate your audience's interpretation of a word cloud by removing certain words during preprocessing steps. This makes choosing your stopwords paramount to a sound visualization which does not mislead. In general, it is best to use word clouds sparingly despite their popularity. If you choose to use word clouds then changing stopwords in an iterative fashion may improve the visual appeal while maintaining a trustworthy intent.

The `wordcloud` library has three interesting word cloud functions. The simplest is named `wordcloud`. It is passed the word terms and a vector of corresponding term frequency and some aesthetics such as orientation, maximum number of words to plot and color. The next function is the `commonality.cloud` function. The commonality cloud adds another informational dimension beyond frequency. A commonality cloud uses multiple documents' terms. The function finds words that are shared among multiple term vectors, and if the words are shared, the function will plot the n-grams. For example, if you want to know the terms that politicians share between parties, you would take words from speeches of two political parties and employ the commonality cloud. The commonality cloud is the conjunction of two groups of words. In contrast, the `comparison.cloud` function will identify dissimilar words among two or more corpora. This function plots the disjunction of two or more corpora. The function constructs a segregated word cloud with each section matching a distinct corpus. Then the function will add labels so that the audience knows what sections belong to each corpus. This type of word cloud can be informative for understanding contrasting language. For example, this type of word cloud can be used to plot different reviews among restaurants. If reviews of a specific restaurant mention long wait times while contrasting reviews of a competitor do not, words like "long" and "wait" will appear in the section of the first restaurant. Figure 3.13 illustrates the differences between word cloud functions. Using `wordcloud` will plot all words while the other functions plot different subsections of the word population.

3.4.1 One Corpus Word Clouds

The `wordcloud` function takes a vector of terms, and then a term frequency vector. In earlier code you created a data frame based on the row sums of a TDM and term names. This was used to construct the frequency bar plot. This data frame is called `freq.df`. As a refresher you may want to look at the first six terms and frequencies. The code below loads the word cloud library and executes the head function to view the data frame made earlier.

```
library(wordcloud)
head(freq.df)
```

The code snippet below will create a word cloud with a maximum number of terms equal to 100. You can also specify the minimum frequency threshold as well. The code also specifies `grey50` and `darkred`. You can change the colors

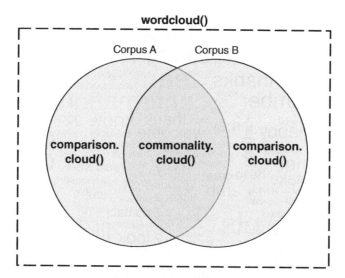

Figure 3.13 A representation of the three word cloud functions from the `wordcloud` package.

manually or, as you will learn later, you can use a predefined color palette. The resulting word cloud is shown in Figure 3.14.

```
wordcloud(freq.df$word,freq.df$frequency, max.words =
100, colors=c('black','darkred'))
```

This word cloud is an example of a non-insightful visual. You already knew the common terms. However, your audience may be able to consume the information more easily from a word cloud than a table or bar plot.

3.4.2 Comparing and Contrasting Corpora in Word Clouds

In order to demonstrate the comparison and commonality clouds, you need to introduce another corpus and change the preprocessing steps slightly. The code should be reminiscent of other code in the book. The custom stopwords now include "sorry," "amp," "delta" and "amazon." The `clean.corpus` function has also been modified. The new custom function is `clean.vec`. The difference is that preprocessing functions are applied to a text vector not a corpus object.

```
library(tm)
library(wordcloud)
tryTolower <- function(x){
  y = NA
```

Figure 3.14 A simple word cloud with 100 words and two colors based on Delta tweets.

```
try_error = tryCatch(tolower(x), error = function(e) e)
if (!inherits(try_error, 'error'))
   y = tolower(x)
return(y)
}
custom.stopwords <- c(stopwords('english'), 'sorry',
'amp', 'delta', 'amazon')
clean.vec<-function(text.vec){
   text.vec <- tryTolower(text.vec)
   text.vec <- removeWords(text.vec, custom.stopwords)
   text.vec <- removePunctuation(text.vec)
   text.vec <- stripWhitespace(text.vec)
   text.vec <- removeNumbers(text.vec)
   return(text.vec)
}
```

Next you import the original delta and new Amazon customer service tweets. The Amazon tweets represent approximately 1000 tweets from @Amazonhelp. If you want to compare more than two text groups you would import them in a similar fashion. Then you would add additional lines of code for each of the following code snippets to incorporate the third or fourth text group into the workflow.

```
amzn<-read.csv('amzn_cs.csv')
delta<-read.csv('oct_delta.csv')
```

Before constructing the word cloud, the specific text vectors have to be cleaned. This is an example of the custom function saving you time because you can apply the function twice without having to rewrite individual code for each text group.

```
amzn.vec<-clean.vec(amzn$text)
delta.vec<-clean.vec(delta$text)
```

At this point, the code diverges from prior examples. Each vector is collapsed into a single document representing the larger Amazon and Delta terms. The resulting object, all, becomes the corpus object with only two documents.

```
amzn.vec <- paste(amzn.vec, collapse=" ")
delta.vec <- paste(delta.vec, collapse=" ")
all <- c(amzn.vec, delta.vec)
corpus <- VCorpus(VectorSource(all))
```

Once you have a clean corpus of two documents, you call TermDocumentMatrix to construct your TDM. The TDM is changed to a matrix and the columns are named "Amazon" and "delta." The column names have to be added in the same order as the corpus was created. This ensures labeling in the visual is correct.

```
tdm <- TermDocumentMatrix(corpus)
tdm.m <- as.matrix(tdm)
colnames(tdm.m) = c("Amazon", "delta")
```

You can examine a portion of the TDM to see the layout and column labels. Table 3.4 shows the portion of the TDM that will print to your console using the code below.

```
tdm.m[3480:3490,]
```

At this point, you could create a commonality cloud. However, the code below will give you more flexibility for coloring the terms in the word cloud. The wordcloud library also loads another package called RColorBrewer. This package has predefined palettes that can save you time compared to constructing your own. To review all pre-constructed color schemes, use the following code. This produces a visual of all palette names with corresponding colors. You can pick one that suits your needs that will be referenced later.

```
display.brewer.all()
```

Table 3.4 The ten terms of the Amazon and Delta TDM.

Term	Amazon	Delta
sonic	1	0
sonijignesh	4	0
sont	2	0
soon	14	16
sooo	0	1
sootawn	0	1
sophiesoph	0	1
soraparuq	0	2
sort	2	0
sorted	5	0
soumojit	1	0

In this example, the `Purples` palette is selected from the `RColorBrewer` package. Originally `Purples` has eight color shades. The next line of code removes the two lightest colors. Removing the lightest colors is important because extremely light colors are hard to read in a cluttered word cloud. You can change from `Purples` to another color name by replacing the text in between quotes to a palette from the previous code's plot. Remember, the names are case sensitive.

```
pal <- brewer.pal(8, "Purples")
pal <- pal[-(1:4)]
```

With your colors selected, you can now construct a commonality cloud. In this example, the more frequent terms will be darker purple, but as you change your colors, this may not be the case. The `commonality.cloud` function accepts the `tdm.m` object directly. The function will identify the words in common and then plot the cloud according to the specified aesthetics. Note that the `pal` object is passed to the colors parameter of the commonality cloud. You could specify colors manually or, in this case, use a predefined color list. Figure 3.15 is the resulting commonality cloud object from the code below.

```
commonality.cloud(tdm.m, max.words=200,
random.order=FALSE,colors=pal)
```

You can compare the contrasting words by calling the `comparison.cloud` function on the same `tdm.m` matrix. The next code selects two colors from the `Dark2` palette. If you are comparing more than two corpora the number of colors will correctly change. The number of colors are distinct because the

Figure 3.15 The words in common between Amazon and Delta customer service tweets.

colors are based on the number of columns in the matrix. For example, if you had three corpora, your TDM would have three columns. The `ncol` in the code below references the total number of columns for selection from the `Dark2` color scheme.

```
comparison.cloud(tdm.m, max.words=200,
          random.order=FALSE,title.size=1.0,
          colors=brewer.pal(ncol(tdm.m),"Dark2"))
```

The code produces a comparison cloud similar to Figure 3.16. The previously defined column names show up as labels in the illustration for each word group. As expected, Delta airline tweets mention flights while Amazon customer service agents tweet about deliveries. This may seem commonsense, but the corpora illustrate the different types of clouds.

3.4.3 Polarized Tag Plot

The problem with the commonality cloud is that words are shown if they are shared among corpora. It does not demonstrate the explicit differences in

Figure 3.16 A comparison cloud showing the contrasting words between Delta and Amazon customer service tweets.

shared words. For example, if a corpus mentions a term once and another corpus mentions it 100 times the term will be plotted in the commonality cloud. In reality, the significant difference between term frequencies may be insightful. The following code constructs a pyramid plot to explore shared words and their corpora differences. Unlike a commonality cloud, the pyramid plot can only be constructed with two corpora. Further, the method to identify common words varies slightly, so some terms may be inconsistent between the commonality, comparison and polarized word clouds. The code in this section is straightforward and can be adjusted if you want to ensure consistency.

First load the `plotrix` package. Next create a subsection of the `tdm.m` using subset. In this code, the logical statement keeps terms that appear more than 0 times in both of the two columns. This is because of the "&" ampersand. The `common.words` matrix contains the same two columns as `tdm.m`, "Amazon" and "Delta". However, instead of containing 4049 terms in the original, the `common.words` matrix has 578.

```
library(plotrix)
common.words <- subset(tdm.m, tdm.m[, 1] > 0 &
tdm.m[, 2] > 0)
```

Examining the tail of the `common.words` object illustrates the effect of the subset function. The tail of the matrix is shown in Table 3.5.

Table 3.5 The tail of the `common.words` matrix.

Terms	Amazon	Delta
working	11	17
wow	2	10
write	1	1
wrong	1	2
yes	4	28
yet	14	3

```
tail(common.words)
```

Notice how all words in the new matrix have a value greater than zero. This matrix is the foundation for your analysis. The next line of code calculates the absolute value of the differences between columns. Calculating the absolute differences treats both columns equally for exploring the differences. You may want to remove the absolute function `abs` that wraps the subtraction. Doing so will change your visualization in favor of exploring the largest term differences for the first column. Leaving the absolute value function will treat all differences the same between columns.

```
difference <- abs(common.words[, 1] - common.words[, 2])
```

The `difference` object is then bound to the common words as a new third column. Then the entire matrix with terms, corpora frequencies and difference columns is ordered by the third column. The rows are ordered in decreasing absolute difference between the words in common between Amazon and Delta.

```
common.words <- cbind(common.words, difference)
common.words <- common.words[order(common.words[, 3],
decreasing = TRUE), ]
```

The matrix is converted to a smaller data frame. The data frame selects the first 25 term values from the Amazon, and Delta columns. The row names are also captured as another column. While redundant, this makes the exact terms easy to comprehend because the data set is smaller. In this example, the `top25.df` object is used to create a pyramid plot.

```
top25.df <- data.frame(x = common.words[1:25, 1],y =
common.words[1:25, 2],
            labels = rownames(common.words[1:25, ]))
```

Lastly you are ready to create the pyramid plot. Using `pyramid.plot` from `plotrix` you pass in the x and y values along with some aesthetics. If you have

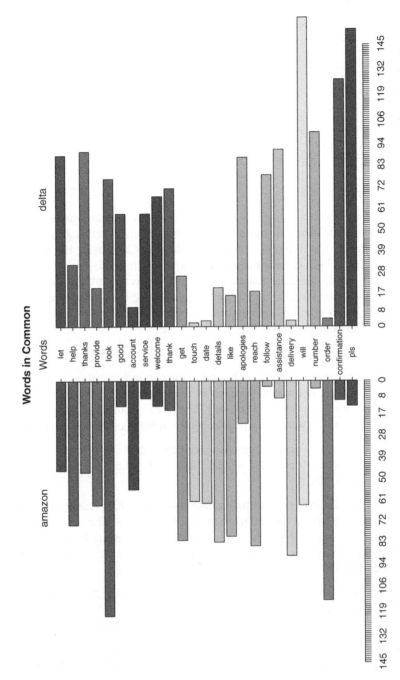

Figure 3.17 An example polarized tag plot showing words in common between corpora. R will plot a larger version for easier viewing.

long words, the bars will cover up the terms making the illustration frustrating for your audience. If that occurs, increase the gap value in the code to increase the distance between horizontal bars. You should also change the `top.labels` parameter to match the headings of your corpora.

```
pyramid.plot(top25.df$x, top25.df$y,
labels = top25.df$labels,
                gap = 14, top.labels = c("Amazon",
"Words", "delta"),
                main = "Words in Common", laxlab = NULL,
raxlab = NULL, unit = NULL)
```

Finally, in Figure 3.17 you see the polarized tag plot. The visualization contains words that are in common but have the 25 largest difference in use. This type of plot may be interesting in reviewing political speeches or in this case, demonstrating that Delta agents are freer to use shorthand "pls" for "please" compared to Amazon agents.

It is a good idea to explore stopwords, and the use of the absolute function earlier when constructing a compelling polarized tag plot. They are less used than traditional word clouds, but may hold more insight in the correct context.

3.5 Summary

In this chapter you learned:

- to visualize simple bar plot of word frequencies
- to find associated words and make a related plot
- to make a basic dendrogram
- to make and improve the aesthetics of a hierarchical dendrogram showing basic word clustering
- how to make a word cloud
- how to compare word frequencies in two corpora and create a comparison cloud
- how to find common words and represent them in a commonality cloud
- how to create a polarized cloud to understand how shared words "gravitate" to one corpus or another
- how to construct a word network quickly

4

Sentiment Scoring

In this chapter, you'll learn

- the definition of sentiment analysis
- a popular academic sentiment framework called Plutchik's wheel of emotion
- what a subjectivity lexicon is
- how to customize a subjectivity lexicon
- the term frequency inverse document frequency weighting (TFIDF)
- Zipf's law
- a basic polarity scoring algorithm
- an archived sentiment library
- a real application of sentiment analysis applied to Airbnb.com Boston area reviews
- the `tidytext` sentiment using an inner join

4.1 What is Sentiment Analysis?

At first thought, sentiment analysis may appear easy: it means distilling an author's emotional intent into distinct classes such as happy, frustrated or surprised. As it turns out, sentiment analysis is very difficult to do well. It borrows from disciplines such as linguistics, psychology and, of course, natural language processing.

Sentiment analysis is the process of extracting an author's emotional intent from text.

Sentiment analysis challenges arise not only from its inter-disciplinary foundation but also from cultural and demographic differences between authors. Another reason is that there are hundreds of related emotional states which are part of the human condition. It is hard to quantify the difference between happy, or elated, or the spectrum of bored to uninterested to interested. In fact, without the author's explicit emotional tone being captured at the time of writing, all sentiment analysis may be undermined by analyst or modeling bias.

Text Mining in Practice with R, First Edition. Ted Kwartler.
© 2017 John Wiley & Sons Ltd. Published 2017 by John Wiley & Sons Ltd.

Further compounding sentiment analysis difficulties may be feature-specific sentiment. This occurs when the topic being written about may have more than one sentiment by feature within the overall topic. For example, a restaurant review on Yelp may state that the prices are great but the food is average. So overall, the review may be decent, but the review itself contains two distinct emotional states (great and average) applied to a specific restaurant feature. Analyzing this type of layered nuanced sentiment is extremely challenging.

There are numerous emotional frameworks that can be used for sentiment analysis. Some are proprietary for commercial applications and others are from academia. A popularized framework was created by Robert Plutchik in the 1980s. Plutchik was a psychologist who created a classification system for emotion. He believed that there are eight evolutionarily created emotions:

1) anger
2) fear
3) sadness
4) disgust
5) surprise
6) anticipation
7) trust
8) joy

He believed that the eight primary emotions have been the basis of survival in humans and animals. As a result, each is foundational to the psyche created over eons. He believed that the eight primary emotions helped to improve survivability over time and were passed on from generation to generation. For example, surprise allowed early humans to make quick assessments as to whether to fight or flee. In this framework, the eight emotions each have a polar opposite. For example, ecstasy is the opposite of grief. To Plutchik, any emotional states outside of these primary eight are amalgamations of the original eight and are therefore subordinate. Lastly, each primary and derivative emotion can be felt to varying degrees. A visual representation of this framework is referred to as Plutchik's wheel of emotion in Figure 4.1.

If you were to create a sentiment model based on Plutchik's framework, then each of the labeled emotions in Figure 4.1 could be a document class in a training set while the document text n-grams could be the independent variables. Then a machine learning algorithm such as Naïve Bayes can be trained and applied to new documents. The end result would be new documents and their corresponding probability for each emotional state and another model for substates. You can start to understand why sentiment analysis is difficult when you consider that Pluthick's approach is just one of many frameworks, and that labeling emotions in the training set is fraught with bias. So it is important that you note methodologies and biases when doing sentiment analysis yourself or when consuming sentiment analysis from others.

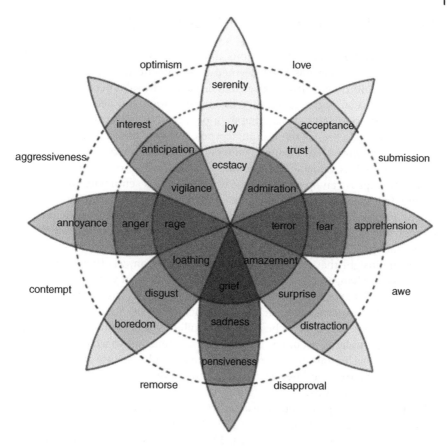

Figure 4.1 Plutchik's wheel of emotion with eight primary emotional states.

Beyond sentiment analysis for emotional states, an easier approach is to merely state whether a document is positive or negative. This is referred to as the *polarity* of a document. Polarity can be more accurate because there are only two distinct classes, and they are easier to disassociate. For example, surprise can be both positive and negative. Positive surprise may be, "I just found out I won the lottery," while negative surprise may be, "I was just hit by a bus." Rather than analyzing the nuanced differences of an emotional state like surprise, polarity of the document is often easier.

This chapter will show the archived `sentiment` package for R that performs basic sentiment analysis. Next, the `qdap` package's polarity function that will also be explained. Finally, the `tidytext` package contains a sentiment scoring approach that will be illustrated.

4.2 Sentiment Scoring: Parlor Trick or Insightful?

In commercial text mining applications and in many academic papers, considerable time and effort has been devoted to sentiment analysis. Despite this effort the results do not always have tangible value. The sales people of some of these commercial organizations try to impress upon the decision-maker a sophisticated approach such as using state-of-the-art deep learning neural nets and truly big data sets as the training corpus. Even so, the value to the enterprise may be limited. For example, understanding a survey respondent's emotional state is less valuable than getting a recommendation about making a change to improve an operation. Some marketers track sentiment over time to attempt to understand the effectiveness of marketing efforts. However, the sentiment scores can be misleading, non-normal or lagged indicators of marketing success, and so the sentiment data should only be accepted with supporting marketing data. In the end, many sentiment analysis vendors do not create an actionable insight that can be used within an operation, whether to improve marketing or change a process. It is less valuable to say "that was negative," than it is to state, "That was negative because of X, Y or Z." The latter requires some subject matter expertise to enrich the sentiment analysis. Still, it is impressive to have sophisticated approaches applied to millions of documents resulting in 80% or better polarity accuracy. But the question remains: to what end?

Despite these limitations, let's embark on an example use case and follow the text mining process outlined in this book to answer a question and thereby reach some conclusions.

Suppose for a moment you have an apartment in Boston that you would like to rent out through the Airbnb.com service. Airbnb is a service for people to list, find and rent lodging, which is used by millions of people throughout the world. You hope to make some extra money by renting your apartment, but you want to make sure that your apartment has the qualities of a good rental.

1) **Define the problem and specific goals.** What property qualities are listed in positive or negative comments?
2) **Identify the text that needs to be collected.** After a stay, an Airbnb renter can leave comments about the property. These comments are public and inform new renters' decisions about the specific property listing. You decide to analyze the comments for properties in Boston.
3) **Organize the text.** The corpus contains 1000 randomly selected Boston Airbnb listings. You will clean and organize the comment frequency matrices.
4) **Extract features.** Once it is organized, you will need to calculate various sentiment and polarity scores.
5) **Analyze.** The sentiment and polarity scores will be used to subset the comments so that you can analyze the terms used distinctly in positive or negative comments.

6) **Reach an insight or recommendation.** By the end of the case study you hope to answer the question from step 1: What property qualities are listed in positive or negative comments? This will help inform you as to whether or not your property has the qualities of a positive Airbnb listing.

4.3 Polarity: Simple Sentiment Scoring

Polarity, the measure of positive or negative intent in a writer's tone, can be calculated by sophisticated or fairly straightforward methods. The qdap library provides a polarity function which is surprisingly accurate and uses basic arithmetic for scoring. The resulting polarity calculation is a number that is negative to represent a negative, zero to represent neutral and positive to represent positive tone. Although the resulting polarity score is easy to understand, it is best to understand the underlying calculation and how to customize it for your specific need by adjusting the subjectivity lexicon.

4.3.1 Subjectivity Lexicons

The polarity function of qdap is based on subjectivity lexicons. A subjectivity lexicon is a list of words associated with a particular emotional state. For example, the words bad, awful and terrible can all reasonably be associated with a negative state. In contrast, perfect and ideal can be connected with a positive state. Researchers at the University of Pittsburgh provide a freely available subjectivity lexicon that is very popular. It contains information on more than 8000 words that have been found to have either a positive or negative polarity. The polarity designation has been captured by various methods and in multiple academic research studies, so it stands to reason that this particular subjectivity lexicon is broadly acceptable. An abbreviated example of the University of Pittsburgh multi-perspective question answering (MQPA) subjectivity lexicon is contained in Table 4.1.

However, the qdap package uses a different subjectivity lexicon for its polarity calculation. Specifically, the lexicon is from research performed by Bing Liu at the University of Illinois at Chicago. This lexicon is slightly smaller, containing approximately 6800 labeled words, but it is based on academic research that has withstood scrutiny. It is important to understand the validity and size of any subjectivity lexicons used in polarity scoring. Errors or biases will have downstream impacts on the output of the analysis.

Armed with either of these or another subjectivity lexicon, you may employ a simple approach of adding up the positive words in a passage and subtracting the negative ones. The net result would yield a number and corresponding positive or negative tone to the passage.

For example, this sentence could be scored using an adding and subtracting method: *"Sentiment analysis in R is good yet challenging."*

Table 4.1 An example subjectivity lexicon from University of Pittsburgh's MQPA Subjectivity Lexicon.

Type	Length	Word	Part of Speech	Stemmed y/n	Polarity
Weak	1	Abundant	Adj	N	Positive
Weak	1	Abundance	Noun	N	Positive
Strong	1	Accede	Verb	Y	Positive
Weak	1	Accept	Verb	Y	Positive
...	...	...	...	...	...
Strong	1	Mar	Verb	Y	Negative
Weak	1	Marginal	Adj	N	Negative
Weak	1	Marginally	Adv	N	Negative
Strong	1	Martyrdom	Noun	N	Negative
...	...	...	...	...	...

The word "good" has a positive polarity, while the word "challenging" has a negative polarity. The two cancel each other out, plus one and negative one equaling zero. So the polarity of this sentence is zero.

However, this is a bit too unsophisticated. You may be thinking that other words outside the subjectivity lexicon could have an effect on the polarity. In particular, these "valence shifter" words, called negation and amplifiers words, likely have an impact on the overall polarity of a passage. A negation word is a word that would actually infer the opposite polarity. Suppose the word "very" in a similar example sentence was changed to "not" as in this sentence: "*Sentiment analysis in python is not good.*"

In this new example, "not" actually should negate the positive intent of the single word "good." The "not good" phrase or token should be considered negative. Rather than just identifying the positive "good" term resulting in a polarity of one, in this case, a more accurate polarity may be closer to negative one. In this example the word "not" is a negation word.

The other type of valence shift is illustrated in the following sentence: "*Sentiment analysis in R is very good.*"

Here the token "very good" likely has a stronger positive tone than the single term "good." So in a simplistic calculation the polarity should be greater than one and possibly approaching two because the "very" indicates a stronger polarity.

As a result of these examples, you should be thinking that any polarity function should somehow account not only for the words in the subjectivity lexicons but also the valence shifters. Luckily the qdap package polarity function does so, and this is explained later in the chapter.

Still another challenge of subjectivity lexicons is that word choice is not universal and is different based on medium and location. For example, in Boston people often use "wicked" in a phrase. The term wicked is largely considered negative elsewhere in the US, yet in Boston the word is sometimes considered good or even an amplifier. Bostonians may say "That's wicked good" when they are expressing "that's really good." Another example of communication medium affecting the word choice occurs in social media. Tweets often contain words like "lol" and "smh." In social media people use "lol" as an abbreviation for the positive "laugh out loud" or "smh" for a disapproving "shaking my head." Yet these terms often do not appear outside of a casual mode of communication and are therefore omitted from many subjectivity lexicons. The polarity function in qdap employs a basic subjectivity lexicon. You should customize the lexicon to suit your text's medium. In the next section you will adjust the lexicons to include specific words.

Why do subjectivity lexicons work?

Despite the shortcomings in these initial examples, subjectivity lexicons are still widely used and are often accurate enough. You may be wondering how such a short list of words from among all known words has the ability to be somewhat accurate. The explanation is based on Zipf's law and the principle of least effort. A small subjectivity lexicon may seem inappropriate because the average person likely has tens of thousands of words in their personal vocabulary, so any list would miss many words known to the author. On top of that, the number of unique words used on any given timeframe varies by gender, age and demographic factors. How can a list of less than 6800 words ever be expected to perform well?

Zipf's law asserts that any word in a document is inversely proportional to its rank when looking at the term frequency. For example, the most frequent word, number one on a frequency list, will occur about twice as often as the second most frequent word, then three times as likely as the third and so on. Table 4.2 shows the first six terms from a frequency distribution used among ~2.5 million tweets mentioning the hashtags #SB50 or #BigGame. Then Figure 4.2 visualizes the top 50 term frequencies from the same corpus. The line chart roughly follows a Zipfian distribution. In both cases, Zipf's law is evident after the first term because the tweets are centered on a particular search term first. After an abbreviation for the football championship, "sb," the most frequent term is "RT," standing for "retweet".

You will notice that the term "RT" appears almost exactly three times as often as the term "to". The term appears approximately four times as often as the term 'a'. Zipf's law occurs in other texts and languages. For example, it occurs in a famous collection of documents called the Brown University Standard Corpus

Table 4.2 The top terms in a word frequency matrix show an expected distribution.

RANK	Word	Frequency	Expected
NA	sb	1,984,423	
1	rt	1,700,564	1700564/1=1,700,564
2	the	1,101,899	1700564/2=850,282
3	to	588,803	1700564/3=566,855
4	A	428,598	1700564/4=425,141
5	for	388,390	1700564/5=340,113

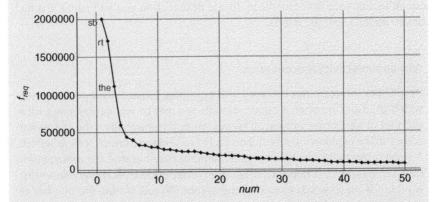

Figure 4.2 Top 50 unique terms from ~2.5million tweets follows Zipf's distribution.

which contains articles published in 1961. In this corpus, there are about 1 million words in total. The term "the" occurs almost 7% (69,971 occurrences) of the time and is followed by "of" which is the second most. True to Zipf's law, the "of" term occurs 3.5% (36,411 occurrences) so the first term is used about twice as often as the second.

One explanation of this linguistic behavior is the principle of least effort. This principle states that humans will choose the path of least resistance and work to minimize effort for a task. Applied to communication theory and library sciences, one understands that the information-seeking audience does not want to exert a lot of effort to understand the message or search for meaning. Once some minimum threshold of understanding has occurred the effort exerted in searching for meaning will decrease or cease altogether. You may have been guilty of this when someone is babbling on about a topic at a cocktail party, yet you already know the topic well. Inevitably you end up not paying full attention and minimizing your effort to listen while not being rude. Your mind starts to wander or you may lose eye contact while you scan the room looking for someone else more interesting to talk too. Likewise, the person writing or speaking wants to minimize their effort to convey meaning. If fewer words will do, then using more is wasted effort. In addition, if using abstruse words increases the effort of the audience, then it is often not wise to use such terms. This is due to the principle of least effort which ensures the audience will tune out, thereby countering the author's intent to convey meaning.

As a result of Zipf's law and the principle of least effort, extremely large subjectivity lexicons may not be needed. It turns out that while humans may know tens of thousands of words, they often revert to using only a few thousand distinct terms when communicating because they want to minimize effort. Due to the predictable distribution of word choice, a subjectivity lexicon that is well researched can safely remove many words. A benefit to having a shorter lexicon is that long word lists take longer to scan and compute, while Zipf's law confirms that many of the words would not be used anyway.

In the previous simplistic examples, some polarity measure was attained, but a more sophisticated approach is needed to boost accuracy. The polarity function of qdap goes beyond one word positive and negative differences in a passage. The qdap approach remains simple yet accounts for negation and amplification words. Also, the function allows you to change the lexicons for specific terms that may or may not be germane to the medium, e.g. "rofl" for "rolling on the floor laughing," which is often used in chat transcripts.

4.3.2 Qdap's Scoring for Positive and Negative Word Choice

The polarity function from qdap is a bit more complex than the previous section's example but is explained easily enough.

1) The polarity function scans for positive and negative words within a subjectivity lexicon.
2) Once a polarity word is found, the function creates a cluster of terms, including the four preceding words and two following words.
3) Within the cluster, neutral words are counted as zero. The positive and negative words that form the basis of the cluster are counted as one and negative one respectively. The remaining non-neutral and non-polarity words are therefore considered valence shifters. These valence shifters are given a weight to amplify or detract from the original polar word. The default value is 0.8. So amplifiers add 0.8 while negating words subtract 0.8.
4) All of the values in the word cluster are summed to create a grand total of the polarity with amplification or negation effects.
5) The grand total of positive, negative, amplifying and negating words with their specific weights is then divided by the square root of all words in the passage. This is helps to measure the density of the keywords.

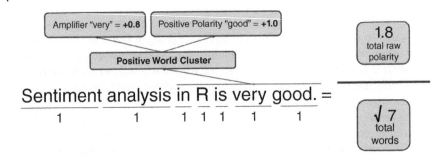

Figure 4.3 Qdap's polarity function equals 0.68 on this single sentence.

Although this five-step process may seem complicated, Figure 4.3 illustrates it graphically using a previous example.

Now that you have an understanding of the polarity function in action you may want to incorporate custom words into the original subjectivity lexicon. Suppose as you have gained subject matter expertise in Airbnb reviews so that you believe authors use terms like "rofl" and "lol." These two terms are not included in qdap's polarity function dictionary. As a result, you will need to append these terms to the list of positive words.

First create a vector of the new positive terms, called new.pos. You add the terms rofl and lol but you can add more simply by using the comma and quotes within the parentheses.

```
library(qdap)
new.pos<-c('rofl','lol')
```

Most likely you will need to retain the original positive words rather than replace them altogether. The basic subjectivity lexicon is held in an object loaded within qdap called key.pol. The key.pol object has both positive and negative terms, but you only need the positive ones to add to. Using the subset function, you are able to retain only the terms in the original key.pol lexicon that have a polarity value equal to one.

```
old.pos<-subset(as.data.frame(key.pol),key.pol$y==1)
```

The code below concatenates the new positive terms with all terms of the old positive lexicon into a new object called all.pos.

```
all.pos<-c(new.pos,old.pos[,1])
```

You are only half done with revising the original polarity lexicon. Now you have to adjust the negative portion in a similar manner. Suppose that your research has determined the words "kappa" and "meh" to be negative, and you would like to include them. It turns out that kappa is a term used among gamers to denote some amount of negative sarcasm. The other term, meh, is more broadly used in the context of being unenthusiastic or apathetic. In any case, you need to create an object with both, called new.neg below.

```
new.neg<-c('kappa','meh')
```

As before, you need to append these new negative terms to the original ones. In order to do so, you create `old.neg` using the subset function. However, this time you are only concerned with terms that have a negative one polarity, so the code changes slightly.

```
old.neg<-subset(as.data.frame(key.pol),key.pol$y==-1)
```

To finish the negative terms, you combine the new negative terms with the old negative terms using the code below.

```
all.neg<-c(new.neg,old.neg[,1])
```

Lastly you need to create a sentiment frame to replace the original. The polarity function refers to this special data frame class when doing its calculation. Here you are creating a new object called `all.polarity` using the `sentment_frame` function. The function needs the vector of all positive terms, all negative terms which you made earlier, along with the corresponding weights to assign to each type of word.

```
all.polarity<-sentiment_frame(all.pos,all.neg,1,-1)
```

Now that you have customized the polarity function's reference words, you can apply it. Consider the short examples here:

> *"ROFL, look at that!"*
> *"Whatever you say. Kappa."*

To invoke the new lexicon, you have to specify it as shown below on the example sentences. The second parameter of the polarity function explicitly redirects to the new sentiment frame containing the original and new subjectivity words. The output of the code below follows in Table 4.3.

```
polarity('ROFL, look at that!',polarity.frame
=all.polarity)
```

In Table 4.3, the first column says "all" because there is no grouping variable, as could be the case with author or date. The second is how many sentences were detected. Next is the total number of words, followed by the average polarity score. The last two columns would be populated if there were more than one sentence.

Table 4.3 The polarity function output

All	Total sentences	Total words	Avg. polarity	Standard dev. polarity	Standard mean polarity
all	1	4	0.5	NA	NA

Table 4.4 Polarity output with the custom and non-custom subjectivity lexicon.

All	Total sentences	Total words	Avg. polarity	Standard dev. polarity	Standard mean polarity
All	1	4	−0.5	NA	NA
All	1	4	0	NA	NA

With the new lexicon in place, "ROFL" was identified as positive, with all other words being neutral. The polarity of this sentence is 0.5. This is because "ROFL" counts as one with three other words being neutral. The one is divided by the square root of the total number of words, 4. In the end, one divided by the square root of four is 0.5.

If you do not specify the custom polarity frame in the code as shown next, the function will revert back to its original less customized polarity words. Without the customized polarity frame the polarity calculation is a neutral 0 because "ROFL" was not found.

```
polarity('ROFL, look at that!')
```

The same behavior occurs with the sentence containing "kappa." In the first line of code the customized polarity frame will calculate a −0.5 score. Using the original polarity frame in the second code line will produce a neutral 0. The presence of "kappa" is identified as a polarized word in the first line. Table 4.4 shows the outputs, side by side, for comparison in the respective order of the code.

```
polarity('whatever you say, kappa.', polarity.frame =
all.polarity)
polarity('whatever you say, kappa.')
```

4.3.3 Revisiting Word Clouds – Sentiment Word Clouds

By now you should have a decent understanding of how a basic polarity function can score a passage and how to customize it for your specific purpose. You can add this dimension of analysis when performing a wordcloud function in the hope of understanding a new insight. In review from the previous chapter you can apply the word cloud package functions to the words in single corpus, or the disjunction or conjunction of multiple corpora. This is represented again in Figure 4.4 originally from Chapter 3.

Once you calculate the polarity for documents in a corpus, you can use this dimension to subset it. You are artificially creating two corpora to visually examine from the single corpus. Creating word clouds with this methodology

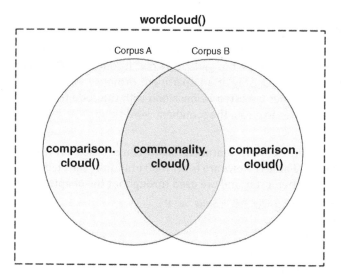

Figure 4.4 The original word cloud functions applied to various corpora.

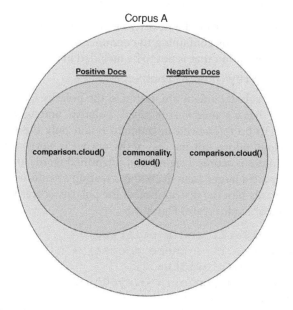

Figure 4.5 Polarity based subsections can be used to create different corpora for word clouds.

will show what distinctive words are used only for positive versus negative posts and which other words are shared, thereby having a mixed sentiment. This is visually represented in Figure 4.5.

Getting the Data

Please navigate to www.tedkwartler.com and follow the download link. For this analysis, please download "bos_airbnb_1k.csv". It contains 1000 Airbnb reviews related to a particular stay at an apartment or house in Boston. The global options and libraries are listed below along with the code to create the initial data frame needed to create the sentiment word cloud.

Let's create an example sentiment word cloud based on a real corpus of online reviews. These reviews are from 1000 randomly selected Boston Airbnb comments left after a stay and are used throughout the chapter.

```
options(stringsAsFactors = F)
library(tm)
library(qdap)
library(wordcloud)
library(ggplot2)
library(ggthemes)
bos.airbnb<-read.csv('bos_airbnb_1k.csv')
```

The next step is to calculate the polarity of each comment. Here polarity is only applied to the vector containing the comments because the data frame contains other information that is not to be included in the score.

```
bos.pol<-polarity(bos.airbnb$comments)
```

The scores for each document are nested in the polarity output `bos.pol`. The element "all" has a scored vector called "polarity" among other information. Both need to be referenced in order to retain only the scores for this specific word cloud effort, but the rest of the output is worth exploring. The `bos.pol` list contains scores, original text, identified words and other information that may aid a larger analysis. Before creating a sentiment-based word cloud, you can easily plot the distribution of the polarity scores using the code below. The `ggplot` code creates Figure 4.6.

```
ggplot(bos.pol$all, aes(x=polarity,
    y=..density..)) + theme_gdocs() +
  geom_histogram(binwidth=.25,
    fill="darkred",colour="grey60", size=.2) +
  geom_density(size=.75)
```

Now that the polarity scores have been calculated, you can append part of it to the original data frame. The next code appends just the polarity scores from the `bos.pol` object to the original `bos.airbnb` data frame as a column alongside the original text.

```
bos.airbnb$polarity<-scale(bos.pol$all$polarity)
```

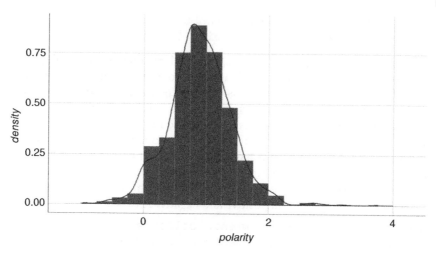

Figure 4.6 Histogram created by ggplot code – notice that the polarity distribution is not centered at zero.

Notice that the scale function is being applied to the polarity output. When doing exploratory data analysis like Figure 4.6, you may notice that the polarity scores are not centered at 0. As shown in Figure 4.6, the mean of the original polarity scores is 0.90. This means that on average each review has at least a single positive word in it. This often occurs in reviews and market research. Despite being anonymous, there is a social norm for people to be nice, acknowledge effort or find at least something to be positive about. This is called response bias and results in a "grade inflation," where people are mixing positive words alongside negative. Scaling the polarity score vector moves the average to zero. It is a good idea to use the scale function and also rerun the analysis without it, to understand the scaling effect on outcomes.

To create the sentiment-based word cloud, you need to create a subset of the original data. This will give you only the documents that are positive or negative. To do so, you will apply the subset function as shown in the code. There are some drawbacks to this approach worth noting. First, the subset will remove the vast majority of documents if most documents score a neutral zero. Second, simply segregating based on greater than or less than 0 means that the degree of polarity is not captured. Another drawback occurs if you want to understand the words in the neutral subsection. If needed, you can add a third object where polarity equals zero. To capture stronger or weaker polarity subsections, you could use additional non-zero subset parameters. One approach may be to quartile the polarity score and use those as the cut-off thresholds. That said, for this example you are simply using greater than or less than zero as your cut-off.

```
pos.comments<-subset(bos.airbnb$comments,
            bos.airbnb$polarity>0)
```

```
neg.comments<-subset(bos.airbnb$comments,
                  bos.airbnb$polarity<0)
```

These two objects represent the corpora for the word clouds. At this point, each contains many hundreds of individual comments, but for this analysis the subsections of reviews should be considered in their entirety. You need to collapse the pos.comments and neg.comments into two distinct documents. To do so use the paste command along with collapse. Then each of these two large comment documents are combined into a single character vector called all.terms. This vector has only two distinct components instead of the hundreds of individual comments in the pos.terms and neg.terms. Finally this is passed to a VCorpus function to become the final corpus with two documents containing all positive and negative reviews.

```
pos.terms<-paste(pos.comments,collapse = " ")
neg.terms<-paste(neg.comments,collapse = " ")
all.terms<-c(pos.terms,neg.terms)
all.corpus<-VCorpus(VectorSource(all.terms))
```

Rather than review all of the preprocessing steps that can be done on a corpus, the code below only performs cursory preprocessing steps. When doing this analysis for real, you would likely create a customized function, as was shown in an earlier chapter, and then apply it to the corpus. The code below creates a TDM using common preprocessing steps. Beyond standard preprocessing, the TDM was constructed by changing the term weight. Changing the term weight is a new parameter that was not done previously. Specifically, the weighting was changed to the term frequency inverse document frequency (TFIDF). As you build this or other bag of words method TDMs, it is good to explore the impact of the matrix weighting.

```
all.tdm<-TermDocumentMatrix(all.corpus,
control=list(weighting=weightTfIdf, removePunctuation =
TRUE,stopwords=stopwords(kind='en')))
```

What is TFIDF?

The TFIDF is the product of the term frequency (TF) and the inverse document frequency (IDF). Hence the TFIDF acronym. Instead of simple term frequency, the TFIDF value increases with the term occurrence but is offset by the overall frequency of the word in the corpus. The offsetting effect helps remove commonly occurring terms that may not yield much information. From a common sense perspective, if a term appears often, it must be important, represented in frequency. However, if it appears in all documents, it is likely not that insightful or informational. This book mentions "text" and "mining" often, so both are

Table 4.5 A portion of the TDM using frequency count

Terms	Positive reviews	Negative reviews
Boss	0	2
Boston	279	259
BostonCambridge	2	0

Table 4.6 A portion of the Term Document Matrix using TFIDF

Terms	Positive reviews	Negative reviews
Boss	0	1.12
Boston	0	0
BostonCambridge	1.13	0

important in some sense. However, every chapter contains these terms, so in reality the value is pretty low compared to other terms.

For another example, the word "Boston" may appear very often in the Airbnb reviews since these are Boston area reviews. A simple term frequency count would probably overemphasize Boston in a word cloud. To combat this, the inverse document frequency is calculated. The inverse document frequency attempts to capture not only the frequency but also the importance of the information the word contains. It is a measure of term frequency times the rarity of the term across documents.

To compare the term frequency and TDIDF values, let's look at a portion of the TDM. Table 4.5 is the portion of the Airbnb term document matrix containing "Boston" using simple frequency count.

In contrast, Table 4.6 contains the same information but with the TFIDF weighting. It is precisely the fact that Boston is so evenly distributed between the two document collections that the TFIDF score is zero. "Boston" is penalized for appearing in so many documents.

Mathematically the TFIDF is defined below.

Term frequency (TF) – counts the number of occurrences of a term in a document. Since words can appear more often simply because documents are long, we can normalize the frequency by dividing by the document length, as shown below.

TF = (term occurrences in a document) / (total unique terms in the document)

Inverse document frequency (IDF) – the log of the total number of documents divided by the number of documents where the term appears

IDF = log(total document in corpus / number of documents with term t in it)

TDIF – the product of TF times IDF
TFIDF= TF*IDF

Figure 4.7 The sentiment word cloud based on a scaled polarity score and a TFIDF weighted TDM.

Once the TDM is constructed, you must switch it to a simple matrix using `as.matrix`. The next step is to label the columns, as was done when constructing a comparison cloud in the last chapter. Both of these steps are performed below.

```
all.tdm.m<-as.matrix(all.tdm)
colnames(all.tdm.m)<-c('positive','negative')
```

Lastly, you call the comparison cloud function specifying the maximum number of words and colors for each polarity subset. The code below is used to construct Figures 4.7 and 4.8. The difference between the visualizations is due to the scaling of the polarity scores in Figure 4.7 and omitting this in Figure 4.8.

```
comparison.cloud(all.tdm.m, max.words=100,
colors=c('darkgreen','darkred'))
```

The word cloud in Figure 4.7 can lead you to an interesting conclusion. In the positive portion of the Airbnb reviews there are many first names such as Alex, Becky, and Erik among others. One could surmise that to have a positive Airbnb experience there must be a direct interaction between host and guest. In contrast, automated postings and dirty rooms lead to negative reviews. Taken collectively, a guest's ideal and expected Airbnb stay is less transactional and more personal but the need for cleanliness is also foundational.

Figure 4.8 The sentiment word cloud based on a polarity score without scaling and a TFIDF weighted TDM.

Without scaling applied to the polarity score, the words themselves change drastically in Figure 4.8. However, one can still draw conclusions, although perhaps less novel about positive and negative Airbnb stays. For positive stays, being within walking distance or having proximity to a subway station are important. Not surprisingly if the room hygiene, condition, safety or proximity to dumpsters is mentioned, the review is negative.

4.4 Emoticons – Dealing with These Perplexing Clues

Emoticons are the combinations of punctuation marks, computer accepted symbols or more recently small images called emoji that are meant to convey information. In a sense, emoticons are shorthand for thoughts and feelings rather than typing out complete words and phrases. An example of a punctuation-based emoticon is ":P" which is meant to be a face with a tongue sticking out. Another example is ":-)" representing a simple smiley face with a nose. Emoticons have gained in popularity and continue to evolve. An outgrowth of the punctuation or symbol based emoticons is the use of emoji, which are cartoon-like images instead of normal keyboard characters used to convey meaning. Emoji change fairly rapidly and vary by medium. For example, many smart phone updates also periodically update the emoji as part of the messaging application. Emoji differ significantly within chat transcripts

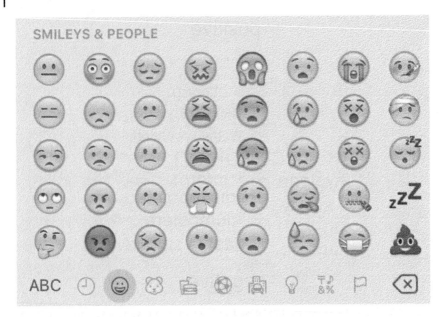

Figure 4.9 some common image based emoji used for smart phone messaging.

compared to smartphone emoji, especially for specialized websites like twitch. com. Figure 4.9 contains just a selection of the smiley face emoji used in a smartphone messaging application. For comparison, Table 4.7 contains common punctuation mark and symbol-based emoticons.

The constant evolution of emoticons and emoji, and the differences between communication platforms make tracking these symbols for sentiment analysis

Table 4.7 A small sample of the hundreds of punctuation and symbol based emoticons.

Emoticon	Meaning	Type
:-)	Happy face	Punctuation
:-D	Laughing face	Punctuation
:(	Frowning face	Punctuation
;)	Winking happy face	Punctuation
ಠ_ಠ	Disapproval look	Punctuation
♥	Heart	Unicode symbol
✂	Cut	Unicode symbol
☺	Smiley face	Unicode symbol

challenging. This section shows how to deal with some common emoticons, but you will need to refine and update them as the emoticons themselves vary.

4.4.1 Symbol-Based Emoticons Native to R

R can interpret some older symbol-based emoticons using the corresponding Unicode. Unicode is a universal manner of displaying text used among computers. It is often represented as "U+2764." This specifies Unicode and character number 2764. However, if you create an object using "U+2764," R will interpret it literally. It does not get translated to its corresponding Unicode character and is instead kept as the specific character string. R needs to have the Unicode "escaped" to signify that the code is representative of a Unicode character. To escape the Unicode use "\" to begin the code and drop the plus sign as shown here.

```
"\U2764"
```

R will print out the result which is actually a heart represented in the console as "[1] "❤"." Many of the older emoticons primarily from word processor "wingdings" can actually be printed and interpreted correctly with this methodology in an R console. Table 4.8 shows some basic example Unicode symbols that R can interpret natively. Using this method, R can print some but not all Unicode characters.

Fixing the Unicode symbol-based emoticons is fairly straightforward. Consider the following sentences:

"I am 😎. I ❤ ice cream."

The dark smile face and heart represent happy and love respectively. As part of the preprocessing for this passage or any other which may include a significant amount of emoticons, it makes sense to change the emoticons to normal character strings. Borrowing from qdap, the mgsub function is helpful for multiple substitutions. The code below searches for the Unicode patterns and replaces them with the symbol's intent. The first object is a character vector for the Unicode equivalent for the smile face and heart. The next object is a

Table 4.8 Example native emoticons expressed as Unicode and byte strings in R.

Emoticon	Description	R Unicode
❤	Heart or love	\U2764
☺	Smiley face	\U263A
☹	Frowning face	\U2639
☻	Dark smile face	\U263B
✌	Peace sign	\U270C

character vector of replacements. Lastly `mgsub` is applied to the text object, which in this case are the example sentences.

```
patterns<-c('\U263B','\U2764')
replacements<-c('happy','love')
mgsub(patterns,replacements,text)
```

The final sentences have the replacements: "I am happy. I love ice cream."

Once the emoticons are in a character string, the symbols or more specifically the substituted words can be calculated as part of sentiment or polarity scoring. If you expect to encounter many of these types of characters, you will need to include more patterns and replacements in the above code. An online search for Unicode emoticons will provide lists that you can incorporate into the example here.

4.4.2 Punctuation Based Emoticons

Another way to convey meaning through symbols is to use punctuation and keyboard letters. Once again, expressing emotion using keyboard characters is dynamic and varied. Some of the popular facial related punctuation emoticons are given in Table 4.9.

Qdap has a built-in dictionary of 81 emoticons and their meanings. To review the emoticons, load the emoticon data frame using the code below after qdap is loaded.
`data(emoticon)`

Table 4.9 Common punctuation based emoticons

Emoticon	Description	Emoticon	Description
:)	Smiley face	:-!	Foot in mouth
:-)	Smiley with nose	:-D	Laughter
:(	Frowney face	:@	Exclamation "What?!"
:-(	Sad with nose	:-0	Yell
:-----)	Liar (long nose)	:-@	Angry
:-@	Scream	=^.^=	Cat
:P	Sticking tongue out (raspberry)	O.o	Confused
:-E	Bad teeth	:*)	Drunk smile
>-)	Evil grin	:-O	Surprised
:-\	Shifty	:-$	Confused
(>_<)	Troubled Face	Q_Q	Crying
(-_-)	Shame	¯_(ツ)_/¯	Shrugging whatever

Table 4.10 Pre-constructed punctuation-based emoticon dictionary from qdap.

Meaning	Emoticon
Alien	(.V.)
Angel	O:-)
Angry	X-(
Baby	~:0
Big grin	:-D
Bird	(*v*)

If interested, you can examine the data frame using the head function to print the first six emoticons and their meanings. This will print out the emoticons in Table 4.10.

```
head(emoticon)
```

When you are working on social media channels, it is important that you build out an emoticon lexicon beyond what is standard. You can append other emoticons to the basic data frame using the code below. Simply add new meanings and emoticons in between quotes and separated by commas as shown.

```
meaning<-c('troubled face','crying')
emoticon<-c('(>_<)','Q_Q')
new.emotes<-data.frame(meaning,emoticon)
emoticon<-rbind(emoticon,new.emotes)
```

Once you are comfortable with a punctuation-based emoticon data frame, you can call the mgsub function, also from qdap, to substitute the punctuation to the corresponding meaning. For example, the sentence below contains multiple punctuation-based emoticons that need to be changed.

"Text mining is so much fun :-D. Other tm books make me Q_Q because they have academic examples!"

You can apply mgsub, referencing the emoticon data frame columns and applying them to the text, which is the example sentence.

```
mgsub(emoticon[,2],emoticon[,1],text)
```

After substitutions, the new sentence reads: "Text mining is so much fun Big Grin. Other tm books make me crying because they have academic examples!"

Although not grammatically accurate, the sentence can now be processed and scored using other text mining methods. The information was not lost by simply removing punctuation marks but instead words were substituted for emotional meaning. Thus, it is important to perform this type of substitution *before* any punctuation removal.

4.4.3 Emoji

Emoji are the small pictures or illustrations used throughout modern day communication to convey meaning. Emoji vary in use and type to convey various emotions. For example, an emoji used to convey sarcasm among smart phone messages appears in Figure 4.10. In contrast, Figure 4.11 is a "kappa" emoji used in chat transcripts on gaming sites like www.twitch.com to also convey sarcasm. According to knowyourmeme.com/memes/kappa, the kappa emoticon is used 900,000 times per day on Twitch and is a likeness of an early employee that created the chat client. The kappa emoji popularity has been sustained on the platform because sarcasm is expressed often in the gaming community.

R has difficulty dealing with emoji. The images within text are parsed and expressed incorrectly. If printed to the console, the emoji are shown as one or more black diamonds with a white question mark inside (◆). When you print the emoticons to the console they are not recognizable as Unicode Transformation Format or more specifically UTF-8, which is the format that R uses for character strings. The eight means that R is using 8-bit blocks to represent a single character. Since the emoji are so varied, R merely assigns a question mark because it does not know how to deal with it.

If you view the text containing the emoji, or save the document, the question marks are converted to a string that is unique for each emoticon. For example, '<ed><U+00A0><U+00BD><ed><U+00B8><U+0092>' is the complete string for the sarcastic face in Figure 4.10. The string may appear to be normal UTF-8, but it does not correspond to an expected UTF character. It is merely R's assignment for that emoticon to some characters it can interpret. Since this emoticon encoding is still unknown to R, another conversion is needed. The code below converts a text object from UTF-8 to UTF-8. This may sound odd, but it will ensure that unknown characters are substituted with a "byte". The byte is a hex code that R creates to represent the unknown character. If you are not working with UTF-8, you should change the first and second parameters of the `iconv` function.

```
iconv(text, "UTF-8" , "UTF-8", "byte")
```

Figure 4.10 Smartphone sarcasm emote.

Figure 4.11 Twitch's kappa emoji used for sarcasm.

The hex byte string is unique and therefore can be substituted like other known emoticons. Table 4.11 represents 20 popular Twitter emoji and the corresponding R byte interpretation for Twitter. This can serve as a solid start as you build out an emoji-recoding data frame.

It is important to note that the byte encoding in Table 4.11 was tested on tweets. Other channels or variations of the emoji may be assigned different byte strings. For example, in a browser, the speak no evil monkey is black and white, but it is a color illustration in Twitter. This difference may impact the byte encoding within R. As a result, you should test the byte encoding carefully when setting up a recoding data frame.

To recode a message to account for emoji, you will once again use qdap's mgsub function. For example, consider the following tweet.

Manually dealing with emoji is hard, makes me 😩. Plus doing it one by one makes me 😴. Luckily qdap's mgsub can help 😁

If you use the twitteR package to grab this tweet, R will print the text in a manner shown below in the console.

```
print(tweet)
"Manually dealing with emoji is hard, makes me \xed\
xed. Plus doing it one by one makes me \xed\
xed. Luckily qdap's mgsub can help \xed\xed\
u0084"
```

Table 4.11 Common emoji with Unicode and R byte representations.

	Emoji	Meaning	Unicode	R byte encoding
1		Speak no evil monkey	U+1F64A	<ed><a0><bd><ed><b9><8a>
2		Tears of joy face	U+1F602	<ed><a0><bd><ed><b8><82>
3		Unamused face	U+1F612	<ed><a0><bd><ed><b8><92>
4		Smiley with heart eyes	U+1F60D	<ed><a0><bd><ed><b8><8d>
5		Smiley with smiling eyes	U+1F60A	<ed><a0><bd><ed><b8><8a>
6		OK hand	U+1F44C	<ed><a0><bd><ed><b1><8c>
7		Face blowing a kiss	U+1F618	<ed><a0><bd><ed><b8><98>
8		Thumbs down	U+1F44E	<ed><a0><bd><ed><b1><8e>
9		Loudly crying face	U+1F62D	<ed><a0><bd><ed><b8><ad>
10		Grinning face with smiling eyes	U+1F601	<ed><a0><bd><ed><b8><81>
11		Flushed face	U+1F633	<ed><a0><bd><ed><b8><b3>
12		Thumbs up	U+1F44D	<ed><a0><bd><ed><b1><8d>

	Emoji	Meaning	Unicode	R byte encoding
13		Clapping hands or person raising both hands in celebration	U+1F64C	\<ed>\<a0>\<bd>\<ed>\<b9>\<8c>
14		Person with folded hands or praying	U+1F64F	\<ed>\<a0>\<bd>\<ed>\<b9>\<8f>
15		Pile of "poo"	U+1F4A9	\<ed>\<a0>\<bd>\<ed>\<b2>\<a9>
16		Face with stuck out tongue and winking eye	U+1F61C	\<ed>\<a0>\<bd>\<ed>\<b8>\<9c>
17		Smiling face with open mouth and smiling eyes	U+1F604	\<ed>\<a0>\<bd>\<ed>\<b8>\<84>
18		Sleeping face	U+1F634	\<ed>\<a0>\<bd>\<ed>\<b8>\<b4>
19		Hushed face	U+1F62F	\<ed>\<a0>\<bd>\<ed>\<b8>\<af>
20		Neutral face	U+1F610	\<ed>\<a0>\<bd>\<ed>\<b8>\<90>

Yet if you use the "view" function on the object containing the text, the tweet looks different yet again.

```
view(tweet)
```

```
Manually dealing with emoji is hard, makes me
<ed><U+00A0><U+00BD><ed><U+00B8><U+00AD>. Plus doing
it one by one makes me <ed><U+00A0><U+00BD><ed><U+00B8
><U+00B4>. Luckily qdap's mgsub can help <ed><U+00A0><
U+00BD><ed><U+00B8><U+0084>
```

So this tweet's emoji will need to be reduced to the byte strings using iconv and the code below. The tweet is now in an object emoji.conv which will show the byte encoding alongside the UTF-8 text.

```
emoji.conv<-iconv(final.df$text,"UTF-8" , "UTF-8",
"byte")
```

```
"Manually dealing with emoji is hard,  makes me
<ed><a0><bd><ed><b8><ad>.
Plus doing it one by one makes me
<ed><a0><bd><ed><b8><b4>.  Luckily qdap's
mgsub can help <ed><a0><bd><ed><b8><84>"
```

Using Table 4.11 as a reference, you can create the search patterns using the column named "R byte encoding."

```
emoji.patterns <-c('<ed><a0><bd><ed><b9><8a>','<ed><a0>
<bd><ed><b8><82>',
'<ed><a0><bd><ed><b8><92>','<ed><a0><bd><ed><b8><8d>',
'<ed><a0><bd><ed><b8><8a>','<ed><a0><bd><ed><b1><8c>',
'<ed><a0><bd><ed><b8><98>','<ed><a0><bd><ed><b1><8e>',
'<ed><a0><bd><ed><b8><ad>','<ed><a0><bd><ed><b8><81>',
'<ed><a0><bd><ed><b8><b3>','<ed><a0><bd><ed><b1><8d>',
'<ed><a0><bd><ed><b9><8c>','<ed><a0><bd><ed><b9><8f>',
'<ed><a0><bd><ed><b2><a9>','<ed><a0><bd><ed><b8><9c>',
'<ed><a0><bd><ed><b8><84>','<ed><a0><bd><ed><b8><b4>',
'<ed><a0><bd><ed><b8><af>','<ed><a0><bd><ed><b8><90>')
```

Then you will have to make the replacements vector using column "Meaning" from Table 4.11.

```
emoji.replacements <-c('Speak no evil monkey','Tears
of Joy Face',
'Unamused face','Smiley with Heart Eyes',
'Smiley with Smiling Eyes','OK Hand',
'Face blowing a kiss','Thumbs down',
'Loudly crying face','Grinning Face with Smiling
Eyes',
'Flushed Face','Thumbs up',
'Clapping hands or person raising both hands in
celebration',
'Person with folded hands or praying',
```

```
'Pile of "Poo"','Face with stuck our tongue and wink-
ing eye',
'Smiling face with open mouth and smiling eyes',
'Sleeping face','Hushed face','Neutral Face')
```

Finally, you can apply the msgub function on the emoji.conv tweet object, alswo passing the emoji patterns and replacement objects in. The text now has the emoji replaced with their meanings from Table 4.11.

```
recode.tweet<-mgsub(emoji.patterns,emoji.
replacements,emoji.conv)
"Manually dealing with emoji is hard, makes me Loudly
crying face. Plus doing it one by one makes me
Sleeping face. Luckily qdap's mgsub can help Smiling
face with open mouth and smiling eyes"
```

In practice you do not need to use the print or view functions. You will likely need to expand and test the byte patterns and replacements beyond those in Table 4.11. Once satisfied you can skip to the conversion and mgsub functions to perform the substitution. The mgsub function can be applied to a vector in a data frame such as the Airbnb comments (bos.airbnb$comments) as a preprocessing step prior to creating subsets of the corpus. This is accomplished using the code line below.

```
recode.airbnb<-mgsub(emoji.patterns,emoji.
replacements,bos.airbnb$comments)
```

4.5 R's Archived Sentiment Scoring Library

There was once an R package called sentiment. Its purpose was to classify documents into specific emotional categories similar to Plutchik's wheel of emotion. It has been archived, meaning the package is not actively updated or maintained. As a result, it may not be accurate or even function properly as you update your R installation. The sentiment package uses a Bayesian approach to document classification. You will learn more about document classification later in the book. For the purposes of the sentiment package, the author uses a small subjectivity lexicon with six emotional categories instead of positive or negative. The six categories are anger, disgust, fear, joy, sadness and surprise. The main function in the package calculates a score for each of these emotional states and then selects the specific emotion with the highest score from among the six.

R has other packages that incorporate APIs to perform robust sentiment analysis with more than six categories. This section is illustrative of an approach that was once native to R. Although not active, this section can be useful to understand one approach for sentiment analysis. If you expect to be analyzing specific emotional states, then you may want to explore one of the many API based libraries since they will be more up to date. Further, the next section illustrates a sentiment analysis function in the `tidytext` package. The archived `sentiment` library is shown as an alternative because the `tidytext` package performs more than just sentiment analysis. Rather than a Bayesian approach to sentiment modeling, the `tidytext` package uses various lexicons. Each of the lexicons can be inner-joined to the text for sentiment analysis.

The first step to install the `sentiment` library is to download the source files for both `sentiment` and a dependent package called `Rstem`. These files end in `tar.gz`. You can download the files at:

- https://cran.r-project.org/src/contrib/Archive/sentiment/
- https://cran.r-project.org/src/contrib/Archive/Rstem/

On each webpage, select the most recent files and save the tar.gz files to your local machine. The R code below will install both from your local source rather than the CRAN repository. You will need to change the file path to your specific download of the tar.gz file.

```
install.packages("C:/Users/John_Doe/Desktop/
sentiment_0.2.tar.gz", repos = NULL, type = "source")
install.packages("C:/Users/John_Doe/Desktop/
Rstem_0.4-1.tar.gz", repos = NULL, type = "source")
```

Once the files are unpacked, you can call on the libraries as you would any other package in your local library.

```
library(Rstem)
library(sentiment)
```

Next, examine the emotion lexicon which was loaded with the package. To review the last six emotion words, use the `data` and `tail` functions.

```
data(emotions)
tail(emotions)
```

This will load the data frame into your environment and then print the last six words and corresponding emotion shown in Table 4.12. In total, there are 1541 distinct words that have been classified into anger, disgust, fear, joy, sadness and surprise represented in Table 4.12.

Table 4.12 The last six emotional words in the sentiment lexicon.

Number	Word	Emotion
1536	Yucki	Disgust
1537	Yucky	Disgust
1538	Zeal	Joy
1539	Zealous	Joy
1540	Zest	Joy
1541	Zestfulness	Joy

The `classify_emotion` function acts to get the log likelihood for a document for each of the six emotions. The output is a matrix where each row is a document and each column is the absolute log likelihood of the document expressing one of six emotions. The output data frame contains another column which is the "best fit" and most likely emotion among the six possible. If no words from the subjectivity lexicon are found or all tones are equal, then the last column would contain an NA.

Use the `classify_emotion` function on the Boston Airbnb reviews with the code below. The function is applied only to the vector of Airbnb comments. The function is nested in `as.data.frame` to change the output from a matrix.

```
emotions.df<-as.data.frame(classify_emotion(
                bos.airbnb$comments))
```

The `emotions.df` data frame will have 1000 rows, one per review, and seven columns representing the scores and most likely sentiment for that particular review. Table 4.13 is an abbreviated output of the emotions data frame.

```
head(emotions.df)
```

Table 4.13 The first six Airbnb reviews and associated sentiments data frame.

Comment	Anger	Disgust	Fear	Joy	Sadness	Surprise	Best fit
1	1.468	3.092	2.067	26.286	1.727	7.340	Joy
2	7.340	3.092	2.067	26.286	1.727	2.786	Joy
3	1.468	3.092	2.067	1.025	1.727	7.340	Surprise
4	7.340	3.092	2.067	32.602	7.340	7.340	Joy
5	7.340	3.092	2.067	1.025	1.727	2.786	Anger
6	1.468	3.092	2.067	7.340	1.727	2.786	Joy

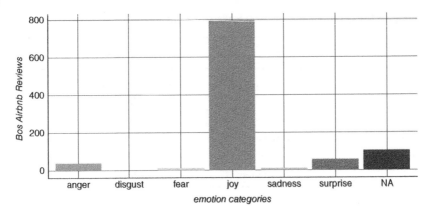

Figure 4.12 The 10k Boston Airbnb reviews skew highly to the emotion joy.

If you review the first six comments, you may disagree with the sentiment analysis outcome. Remember that this is a fairly unsophisticated approach with limited lexicons, and that the package itself is not actively updated. Despite these limitations, the package shows some merit and also helps you to understand one approach to sentiment analysis. An advanced R user could review the function code looking for ways to improve the results and Bayesian model. For now you can make a bar plot of the best fit and compare to the qdap polarity distribution in Figure 4.6. Figure 4.12 uses `ggplot` to construct a frequency count of the best fit sentiment. The code refers to the emotions data frame, and then specifies that the x axis is a count of the emotional categories. The rest of the code identifies aesthetics. In both polarity and sentiment, you see the comments skewing to positive and joy, so both analyses are aligned overall.

```
ggplot(emotions.df, aes(x=BEST_FIT)) +
  geom_bar(aes(y=..count.., fill=BEST_FIT)) +
labs(x="emotion categories",
    y="Bos Airbnb Reviews")+theme_gdocs() +
  theme(legend.position="none")
```

Revisiting the sentiment word cloud, you can now redefine the corpus subsections beyond positive and negative. Although this example is heavily skewed to "joy," other text mining efforts may yield more insights than this simple example. Due to this imbalance, joy shares many words with all other emotional states. The comparison cloud does not plot many words that are distinct to that emotion. However, within the next visual it does appear that walking distance to restaurants leads to a joyful review. In contrast, wanting to be closer and in the neighborhoods of Malden and Somerville leads to a small number of reviews expressing disgust.

First you append the emotional categories to the original review data frame. Instead of using subset with a polarity score, you split the reviews by emotional category to construct Figure 4.13. The rest of the code follows the same

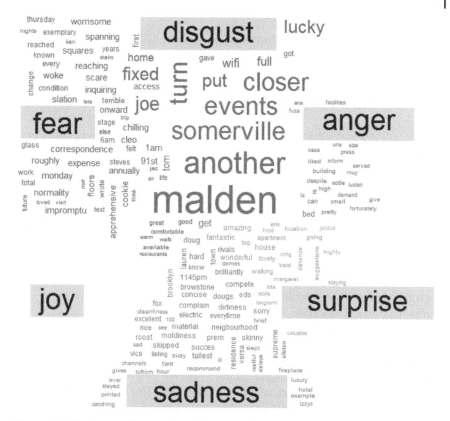

Figure 4.13 A sentiment-based word cloud based on the 10k Boston Airbnb reviews.
Apparently staying in Malden or Somerville leaves people in a state of disgust.

pattern as before but is applied to a list since there are more than two subsets
of the text.

```
bos.airbnb$emotions <-(emotions.df$BEST_FIT)
emotion.reviews <-split(bos.airbnb$comments,
               bos.airbnb$emotions)
emotion.reviews <-lapply(emotion.reviews,
paste,collapse=" ")
emotion.reviews <-do.call(c,emotion.reviews)
emotion.reviews <-VCorpus(VectorSource(
               emotion.reviews))
all.tdm <-TermDocumentMatrix(emotion.reviews,control=
list(weighting=weightTfIdf, removePunctuation = TRUE,
stopwords=stopwords(kind='en')))
```

```
all.tdm.m<-as.matrix(all.tdm)
colnames(all.tdm.m)<-levels(bos.airbnb$emotions)
comparison.cloud(all.tdm.m)
```

4.6   Sentiment the Tidytext Way

As a small diversion from the case study, you will learn sentiment functions from the `tidytext` package. The package is used to perform text mining using the `tidy` format. Rather than continue to work on the Boston Airbnb reviews, this short explanation uses a copyright-free book called *The Wonderful Wizard of Oz*. The book and classic movie continue to be popular, and are about a young girl taken to a distant fantastical land and returning home. The book is in the public domain so it can be downloaded from various online sites including www.tedkwartler.com. This corpus, along with the `tidytext` functions, illustrates a chronological view of sentiment. The Airbnb review data is not a time series data set with periodicity. In contrast, within *Wizard of Oz*, sentiment can be tracked as the story unfolds line by line and thus shows periodicity.

The `tidytext` package contains a data frame called `sentiments`. It can be loaded using the next code. The data frame contains over 23,000 terms from three different subjectivity lexicons. The first column of the data frame includes the individual words. The next column contains the sentiment associated with the words. The sentiment classes include positive, negative, anger, anticipation, disgust, fear, joy, sadness, surprise, trust or NA. The Bing lexicon portion contains only positive or negative in the sentiment column. The AFINN lexicon section contains NA in this column and the NRC subdivision has all sentiments except NA. This is because scoring among the lexicons varies. The third column associates the row with a lexical source. The lexicon options include Bing, AFINN and nrc. The last column contains a numerical score between –5 and 5. These values only occurs for the AFINN lexicon while the other two contain NA. Table 4.14 shows some example rows from the three lexicons for comparison.

As with any sentiment scoring, you should understand its source and methodology. Having the three lexicons in a single data frame makes it useful for quick comparisons. More specifically, the AFINN lexicon has 2476 words and their associated emotion. Fin Arup Nielsen, the AFINN author, is a Danish researcher at a technical university. The second lexicon, containing 6788 terms, is called "bing." It is named for Bing Liu at the University of Illinois, Chicago. This is the same lexicon as is used in the qdap polarity function mentioned earlier. The "nrc" lexicon is the last set of words in the sentiments data frame. The "nrc" data comprises 13,901 words associated to emotions. This data set was labeled using Amazon's Mechanical Turk crowd sourcing service. While the "nrc" lexicon set is large, the biases and problems of crowd sourcing may

Table 4.14 Excerpt from the sentiments data frame.

Word	Sentiment	Lexicon	Score
abhorrent	NA	AFINN	−3
cool	NA	AFINN	1
congenial	positive	bing	NA
enemy	negative	bing	NA
ungrateful	anger	nrc	NA
sectarian	anger	nrc	NA

impact the lexicon's usefulness. When analyzing sentiment with `tidytext`, compare all three lexicons to reach a more generalized conclusion.

Use this code to manually compare and contrast lexicon differences. The code creates three subsets of the original sentiments data frame.

```
library(tidytext) data(sentiments)
afinn<-subset(sentiments,sentiments$lexicon=='AFINN')
bing<-subset(sentiments,sentiments$lexicon=='bing')
nrc<-subset(sentiments,sentiments$lexicon=='nrc')
```

Getting the Data

Please navigate to www.tedkwartler.com and follow the download link. For this analysis, please download "Wizard_Of_Oz.txt". This is a raw text file of the famous book Wizard of Oz.

After downloading the Wizard of Oz text file, load it along with `tidytext`, `dplyr`, `tm` and the `tidyr` packages. `Tidytext` applies tidy data principles to text mining. The `dplyr` library is used for additional data manipulation methods. Of course, by now you should be familiar with the `tm` package! The last package, `tidyr`, has easy functions to manipulate non-tidy data into the tidy format. The last line constructs the `oz` object by reading lines of the text file. The resulting `oz` object is a character vector containing the book.

```
library(tidytext)
library(dplyr)
library(tm)
library(tidyr)
library(ggthemes)
```

```
library(ggplot2)
oz <- readLines("Wizard_Of_Oz.txt")
```

A tidy corpus has a different structure compared to other text mining packages. A tidy corpus has one word per row per section, such as line. A tidy formatted corpus could also have additional columns such as chapter, author or document ID. Also, tidy data format proponents use "%>%", the pipe operator, to pass objects to functions. This is compact, but can be challenging for new R programmers to comprehend.

To begin, construct a DTM using tm principles. The explicit functions here should be familiar since the code mirrors earlier code examples.

```
oz.corp<-VCorpus(VectorSource(oz))
clean.corpus<-function(corpus){
  corpus <- tm_map(corpus, content_transformer(tolower))
  corpus <- tm_map(corpus, removeWords,
stopwords('english'))
  corpus <- tm_map(corpus, removePunctuation)
  corpus <- tm_map(corpus, stripWhitespace)
  corpus <- tm_map(corpus, removeNumbers)
  return(corpus)
}
oz.corp<-clean.corpus(oz.corp)
oz.dtm<-DocumentTermMatrix(oz.corp)
```

Luckily the `tidytext` package provides a simple method for changing a traditional DTM to a tidy format. Apply the `tidy` function directly to the DTM to create a tidy version.

```
oz.tidy<-tidy(oz.dtm)
```

If desired, index a portion of the tidy corpus to examine it. The 6,810th to 6,815th rows from `oz.tidy` are shown in Table 4.15. Notice that the document vector contains the line number containing the word. Unlike qdap's word frequency matrix, the tidy format does not aggregate the words. Instead, the terms are ordered chronologically as they appear in the story. This captures the unfolding story's periodicity as meta information. Lastly, the term vector contains the token, and the count is the number of occurrences in that line.

```
oz.tidy[6810:6815,]
```

Later you will need shared column names for performing an inner join between "term" in `oz.tidy` and "word" in `sentiments`. An inner join compares each row from "Table A" with each row of "Table B" to find all pairs of rows that are shared. In this case, the inner join will identify the words in `oz.tidy` that are contained in the `sentiments` lexicon. The documents or rows

Table 4.15 A portion of the tidy text data frame.

Document	Term	Count
1790	Brought	1
1790	Little	1
1790	Mice	3
1790	Middlesized	1
1790	One	1
1791	Dorothy	1

with a match between the two tables are returned. But first you need to have a shared column, so the code below renames the oz.tidy vectors. Also, change the line number vector to numeric, instead of character, with the second code line. Later, the numeric line number is used to construct the timeline visual.

```
colnames(oz.tidy)<-c('line_number','word','count')
oz.tidy$line_number<-as.numeric(oz.tidy$line_number)
```

(The following code does not use the pipe operator to forward objects to new functions, but code containing the pipe operator is provided at the end of the section for comparison. Using the pipe operator is aligned to tidy data principles, but the less compact code can be more enlightening for learning.)

Apply subset to the sentiments data frame to select the words from the nrc lexicon representing joy. Once you have the subdivided the nrc.joy object, perform an inner join on the oz.tidy table. The inner_join function automatically identifies the shared column "word" to make the join. Finally the count function sums observations by group. This will provide a summarization of all "joy" words found between the two tables. A portion of the result is shown in Table 4.16.

```
nrc.joy <-subset(sentiments,sentiments$lexicon=='nrc'
& sentiments$sentiment=='joy')
joy.words<-inner_join(oz.tidy,nrc.joy)
joy.words<-count(joy.words, word)
```

To construct a polarity timeline, first subset the "bing" words. The code drops sentiments$score. Recall that this column contains integers between –5 and 5 but is only NA for the "bing" lexicon.
```
bing<-subset(sentiments,
        sentiments$lexicon=='bing')[,-4]
```

Table 4.16 The first ten Wizard of Oz "Joy" words.

Word	"n"
Abundance	1
Alive	5
Amuse	2
Amused	1
Approve	1
Art	2
Baby	3
Beam	1
Beautiful	38
Beauty	3

Once again perform an inner join between words in the oz.tidy object and the bing terms.

```
oz.sentiment<-inner_join(oz.tidy,bing)
```

Next use count to summarize the observations. A third parameter is passed into the count function in this example. A new vector called index is created in the new object oz.sentiment. The index vector retains the line information for use later.

```
oz.sentiment<-count(oz.sentiment,sentiment,
index=line_number)
```

Within the tidyr package the spread function is used to spread a key-value pair across multiple columns. First pass in the data frame from count. Then specify the key, the sentiment column. The n column contains the count of positive or negative terms identified. This column is passed into spread as the value. Lastly should there not be any key value pairs so you specify a 0 to be filled in.

```
oz.sentiment<-spread(oz.sentiment,sentiment,n, fill=0)
```

The spread effect is to create a data frame with three columns. Index, containing line information, is retained and the positive and negative words are represented as new columns. This is best illustrated by examining the data frame directly. In rows 15 to 20 you can see that the 55th line in the book contains one negative word and two positive words. In contrast, line 56 contains two positive terms. This section of oz.sentiment is illustrated in Table 4.17.

```
oz.sentiment[15:20,]
```

Table 4.17 The oz.sentiment data frame with key value pairs spread across the polarity term counts.

Index	Negative	Positive
55	1	2
56	0	2
57	1	1
59	0	1
61	0	1
72	0	1

The last two lines of data preparation create two additional vectors. The first, oz.sentiment$polarity, is equal to the positive term count minus the negative term count. The oz.sentiment$pos vector uses an ifelse statement. Logically the function states, "if the polarity vector is 0 or greater then assign "pos" otherwise "neg." This defines a categorical vector for polarity that is positive or negative to be used in a visualization.

```
oz.sentiment$polarity<-oz.sentiment$positive-
  oz.sentiment$negative
oz.sentiment$pos<-ifelse(oz.sentiment$polarity >= 0,
"pos", "neg")
```

The ggplot function is used to create a barplot for each of the text lines. The x value is the index position. The y value defining the height of the bars is the polarity count for the line. Then the pos vector is passed as the color fill for the bars. The next layer of the ggplot adds the bars. Finally theme_gdocs is applied for quick premade aesthetics. The resulting bar plot is represented in Figure 4.14.

```
ggplot(oz.sentiment, aes(x=index, y=polarity,
fill=pos)) +
  geom_bar(stat="identity", position="identity",width=
1)+theme_gdocs()
```

From this view, it may be difficult to understand the general polarity over time, because the bars are visually jagged. Just after line 1000 and before 3000 there are some consistently negative terms used. However, a smoothed visual may be more insightful compared to the choppy bar chart.

To create a smooth line running the epoch of the story utilize geom_smooth when constructing a time series. The geom_smooth layer adds a "smoothed conditional mean" to the plot. The smoothed line aids in identifying patterns amid busy visuals. In this case, the smoothing line is created with a generalized

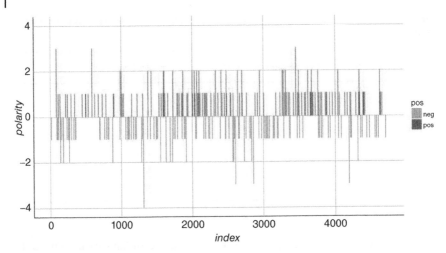

Figure 4.14 Bar plot of polarity as the Wizard of Oz story unfolds.

additive model (GAM). The GAM fits a linear model to the polarity values in `oz.sentiment`. The confidence intervals are estimated based on the fitted model and are also plotted along with the GAM line.

Adding a smoothing function to a `ggplot` is straightforward. After creating the base layer using `ggplot`, add the `stat_smooth` layer and the theme for aesthetics. The code result exemplifies the smoothed Wizard of Oz story arc in Figure 4.15.

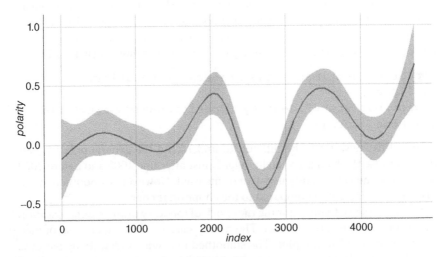

Figure 4.15 Smoothed polarity for the Wizard of Oz.

```
oz.smooth<- ggplot(oz.sentiment, aes(index, polarity))
oz.smooth + stat_smooth()+theme_gdocs()
```

Figure 4.15 provides more context compared to the bar plot in Figure 4.14. For those unfamiliar with the story, Dorothy, the protagonist, is caught in a storm and taken to a fantastical land with witches and creatures. In order to get home she embarks on a journey, picking up friends along the way and seeks the help of a powerful wizard. The smoothed polarity line elucidates her journey. Readers are drawn into the story with fairly neutral language. Around line 2700 the language turns more negative before continually rising to its most positive at the end of the story. In fact, after line 3000 the polarity is completely positive. Without reading the book, one could infer that the story has a happy ending. In contrast, applying this method to a Greek tragedy could change the polarity arc drastically.

In keeping with the `tidytext` package authors' examples, the code below uses the pipe operator. The resulting objects and visuals will be the same. Using the pipe operator means that the code is compact and efficient but hard to understand for new programmers. Compare and contrast the code when you feel comfortable with the `tidytext` sentiment concepts outlined earlier.

This code subsets the sentiments data frame for "nrc" and "joy."

```
nrc.joy <- sentiments %>%
  filter(lexicon == "nrc", sentiment == "joy")
```

The count function provides a table of the words identified in the inner join that mirrors the previous count function.

```
joy.words<-oz.tidy %>%
  inner_join(nrc.joy) %>%
  count(word, sort = TRUE)
```

Select the "bing" lexicon to begin the reconstruction for the Wizard of Oz timeline. The negative sign in the function parameter `select` drops the vector by its name.

```
bing <- sentiments %>%
  filter(lexicon == "bing") %>%
  select(-score)
```

The `oz.sentiment` object is created by forwarding the tidy version of the corpus to the inner join. Next, the object is forwarded to count. Then spread receives the data frame. Mutate is used to add, remove or create new columns in a data frame. Here the mutate function creates polarity with the equation positive minus negative. With tidy data functions column names can be specified by name and without quotes. The last mutate function adds the pos column. Within the function the same `ifelse` statement is presented to create the "pos" or "neg" polarity attributes. The resulting `oz.sentiment` object can be used to create Figures 4.14 and 4.15.

```
oz.sentiment <- oz.tidy %>%
  inner_join(bing) %>%
  count(line_number, index = line_number, sentiment) %>%
  spread(sentiment, n, fill = 0) %>%
  mutate(polarity = positive - negative)%>%
  mutate(pos=ifelse(oz.sentiment$polarity >= 0, "pos",
"neg"))
```

4.7   Airbnb.com Boston Wrap Up

Turning our attention back to the Airbnb case study, you need to complete the text mining project workflow. Using the word cloud visualizations, you can safely answer the questions raised in step 6 of the text mining process outlined earlier in the chapter.

In step 6 of the text mining process outlined earlier, you hoped to answer some basic questions.

6) **Reach an insight or recommendation** – What property qualities are listed in positive or negative comments? Does your apartment have sought-after Airbnb qualities?

Based on the sentiment subsections used to construct various visualizations, if your apartment is clean, has nothing broken, the dumpsters are not close and the apartment is within walking distance to a subway station, you can be reasonably assured of a positive review. Further, you can bolster your guest reviews by providing a personalized interaction in some way. Within the limited analysis, the Airbnb community uses language inferring that personal interactions are also positive.

4.8   Summary

In this chapter, you learned:

- how to apply polarity scores and sentiment categories to a real use case
- a popular academic sentiment framework called Plutchik's wheel of emotion
- what a subjectivity lexicon is
- how to customize a subjectivity lexicon

- the term frequency inverse document frequency weighting (TFIDF)
- Zipf's law
- a basic polarity scoring algorithm
- an archived sentiment library
- a real application of sentiment analysis applied to Airbnb.com Boston area reviews
- the `tidytext` sentiment using an inner join

5

Hidden Structures: Clustering, String Distance, Text Vectors and Topic Modeling

In this chapter, you'll learn

- how to perform k-means clustering
- how to perform k-mediod clustering
- to use the `StringDist` library with `Hclust`
- what LDA is
- LDA topic modeling using LDA and LDAvis
- other topic modeling packages
- use `word2Vec` to get text vector calculations
- make a compelling treemap visualization.

5.1 What is clustering?

In machine learning there are two methodologies. The first is called supervised learning and the second is unsupervised learning. In supervised learning each observation consists of independent attributes with a final outcome or dependent variable for the observation. For example, pretend you owned an ice cream shop and wanted to predict how many scoops you were going to sell. So you decide to create a supervised model for this prediction. The data you collect is for each historical day with attributes such as day of the week, temperature, month of the year along with the outcome variable scoops. Once this is collected and set up as such, you could then apply any number of supervised learning algorithms such as random forest. In contrast, unsupervised techniques do not have dependent variables for each observation. Once you have your ice cream data organized by day, you could apply an unsupervised approach to find the underlying pattern within the data. The goal would be to identify similar days from the data. The difference is that you did not explicitly tell your algorithm to predict anything, you leave the algorithm to search on its own thereby being unsupervised. The underlying pattern is used to partition the data into clusters or subsections of days for your shop. The end result of

Text Mining in Practice with R, First Edition. Ted Kwartler.

your unsupervised approach may be a clustering of days with high scoop sales with attributes like Saturdays with temperatures above 85° and another cluster for Tuesdays with a cooler 50° in October. This chapter covers common unsupervised approaches applied to text and a subsequent chapter illustrates supervised modeling with text.

Applying clustering methods helps to organize individual documents into groups of similar documents. For example, a clustering technique for newspaper articles might partition each article into clusters representing business, sport and politics among other clusters. Document clustering can sometimes be used to extract broad topics from a large group of documents and also to aid in information retrieval systems.

Clustering techniques can help elucidate the topics within a broad corpus without reading the documents. As a marketer you can apply these methods to forum posts or online reviews to understand the broad topics that reviewers find meaningful. This can help inform product innovation and sources of customer angst. If you were in a healthcare compliance role, clustering techniques might help to find outlier and possibly fraudulent expense descriptions.

Document clustering is also used for fast information retrieval because documents can be automatically tagged without human intervention. In another business related example, this means that a database of millions of articles does not need to be read and tagged for finding later. Instead a clustering algorithm can be applied to understand what each article mentions and then that tag is stored as an article attribute. For example, a researcher might want to get a specific group of business articles mentioning Atlanta. The system could subset to only the cluster of business articles, and then perform a smaller keyword search on Atlanta. This is almost certainly faster than searching many millions of articles for both business and Atlanta.

This chapter aims to give you numerous approaches to clustering documents and provide you with exposure to topic modeling. Each has different approaches and popularity. While the technological method varies significantly the differences often boil down to speed versus accuracy. Thus it is important to apply multiple approaches balancing speed and accuracy for your particular application needs.

5.1.1 K-Means Clustering

One method of clustering documents is called k-means clustering. It is a simple approach and relatively easy to comprehend so it is a good place to start. Within the context of documents, k-means clustering seeks to partition each document into a number of predefined "k" clusters which should be similar in the terms used. K-means clustering follows the high-level steps below.

1) Choose the number of clusters represented by "k." The number of clusters must be chosen before any partitioning.

2) For every cluster k, a random point is selected as the "centroid" by the algorithm.
3) Each document is assigned to the closest centroid.
4) Once all documents are assigned, the sum of all distances to the centroid is calculated. Then the centroids are moved to the average distance sum of all assigned documents from the original randomly chosen location. When the centroids move, the clusters will gain and lose documents. The algorithm will continually redefine cluster boundaries and adjust mean centroids as long as sum of distances can be minimized by moving the centroid.
5) The algorithm will stop recomputing centroids when the sum of cluster distances cannot be reduced further. This would mean that all documents stay in their cluster.

Consider the following nonsensical documents which we can use in a simple illustration of k-means. Table 5.1 shows the documents that are made of only two terms with varying frequency.

Reviewing this corpus, you can calculate the saturation of terms "text" and "mining." For each document, each term is divided by the entire length or number of words. The results are shown in Table 5.2.

Once this is done, you can plot the documents in a two-dimensional space. Of course, with real documents they exist in hyperspace but that would not be possible to illustrate. Figure 5.1 shows each document plotted by the corresponding "text" and "mining" value.

To begin a k-means clustering algorithm on this corpus you decide to choose two clusters. Next k-means will assign a random "centroid" and assign documents to the closest cluster. In Figure 5.2, the solid black dots represent arbitrary centroids to begin the partitioning. The elliptical lines are the clusters of documents themselves.

You can probably see that the randomly assigned centroids are not correct. Simply looking at the visual, you can guess that documents 3 and 5 should be a cluster while 1, 2 and 4 are more similar. The upper left cluster includes document 3 because the randomly assigned centroid is closer than the other

Table 5.1 A sample corpus of documents only containing two terms.

Document	Text
D1	Text text text text mining
D2	Text text text mining
D3	Text mining mining mining mining
D4	Text text text mining mining
D5	Mining mining mining

Table 5.2 Simple term frequency for an example corpus.

Document	Term = Text	Term = Mining
D1	0.80	0.20
D2	0.75	0.25
D3	0.25	0.75
D4	0.60	0.40
D5	0.00	1.00

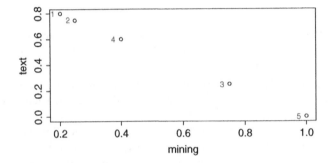

Figure 5.1 The five example documents in two-dimensional space.

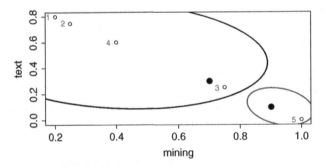

Figure 5.2 The added centroids and partitions grouping the documents.

centroid. At this point, the k-means algorithm will move the centroids to minimize the sum of distances. An analogy to this is that all documents have a gravitational pull like planets. The center of the document solar system is the point at which all "gravity" pulls are equal. Figure 5.3 shows the transition of the centroids. The old centroid coordinates are marked with an "X" and the arrow shows the movement to a new spot.

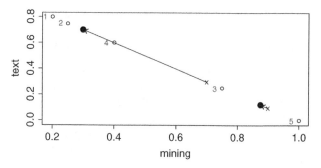

Figure 5.3 Centroid movement to equalizing the "gravitational pull" from the documents, thereby minimizing distances.

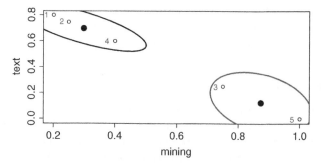

Figure 5.4 The final k-means partition with the correct document clusters.

The last step is to redefine the partition boundaries. With the movement of the centroids, the documents cluster as expected between 3 and 5 and the rest. This is shown in Figure 5.4.

You can now apply the k-means clustering approaches in the following code. An example HR case study is outlined within the context of the six-step text mining process from Chapter 1. For the case study, assume that you work in an HR function at a call center. Call centers often have high employee turnover as the jobs can be unsatisfying, low wage and repetitive. Constantly working on talent acquisition due to high turnover is expensive in call centers. Thus you decide to understand employee characteristics based on their prior work experience. As an HR professional with a data driven mentality you wonder if clustering on employee prior work experience will lead you to identify attributes of employees that are successful in call centers. The hope is to find employee resume clusters that have stayed longer than one year linked by common resume terms. These terms may give you insights for candidate attributes to focus on when interviewing new call center representatives. If you were performing this analysis for real you would likely need more resumes and think explicitly about your organization's needs to make an appropriate selection, e.g. using performance reviews instead of tenure.

Reviewing the six-step text mining process in the context of this case study

1) **Problem definition and specific goals.** Can you search for common traits among employees who have lasted more than a year from their resumes?
2) **Identify the text that needs to be collected.** You have 50 resumes from employees who have stayed more than a year at your organization. From each of the resumes, you select the most recent employment description.
3) **Text organization.** You will organize your text into a DTM with TF-IDF weighting.
4) **Feature Extraction.** You will perform a k-means clustering algorithm.
5) **Analysis.** You will analyze the cluster membership of your corpus and identify prototypical terms.
6) **Reach an insight or recommendation.** The goal is to identify three distinct terms in employee clusters that may help recruiters identify candidates likely to stay more than a year, based on past experiences.

Getting the Data

Please navigate to www.tedkwartler.com and follow the download link. For this analysis, please download "1yr_plus_final4.csv." Keep in mind due to the nature of personal information the work experiences have been modified and disassociated from the entire resume. They have been constructed from randomized portions of public resumes posted online. The work experiences are then organized into a data frame for the analysis.

The code section below is provided without explanation because the concepts and functions were presented earlier in the book. This represents a basic foundation to execute a k-means clustering algorithm and utilizes minimal text cleaning. The code libraries include tm for text mining, flexible procedures for clustering (fpc) and cluster.

```
options(stringsAsFactors = F)
set.seed(1234)
library(skmeans)
library(tm)
library(clue)
library(cluster)
library(fpc)
library(clue)
library(wordcloud)
```

```
clean.corpus <- function(corpus){
  corpus <- tm_map(corpus, removePunctuation)
  corpus <- tm_map(corpus, removeNumbers)
  corpus <- tm_map(corpus, content_transformer(tolower))
  corpus <- tm_map(corpus, removeWords,
c(stopwords("en"), "customer",
"service","customers","calls"))
  return(corpus)
}

wk.exp<-read.csv('1yr_plus_final4.csv', header=T)
wk.source <- VCorpus(VectorSource(wk.exp$text))
wk.corpus<-clean.corpus(wk.source)
wk.dtm<-DocumentTermMatrix(wk.corpus,
control=list(weighting= weightTfIdf))
```

You can now start to perform a cluster analysis using the work experiences constructed as a DTM called `wk.dtm`. You should normalize the term vectors because the DTM is made of continuous numeric features and the algorithm relies on an average of these numeric attributes. This is accomplished with the next code using the `scale` function to create a scaled object called.

```
wk.dtm.s<-scale(wk.dtm,scale=T)
```

The k-means clustering function is installed as part of the base R installation. It is part of the base stats package. There are more in-depth explanations of the k-means functions in books devoted to machine learning so here the code simply calls the function on the scaled matrix and specifies that three clusters should be partitioned.

```
wk.clusters<-kmeans(wk.dtm.s,3)
```

The object `wk.clusters` represents the work experience cluster output. In it each document is assigned a cluster, and some evaluation metrics are returned. An easy place to start evaluating your results is by plotting the total number of documents assigned to each of the three clusters. The returned k-means object has an element that captures this information. A basic plot function can be used to show the outcome. Figure 5.5 shows a very unbalanced document cluster from `wk.clusters`.

```
barplot(wk.clusters$size, main='k-means')
```

Next you may want to plot the clusters to get a sense of the partition separation. This is accomplished with `plotcluster`. In this example, the `plotcluster` function uses the `cmdscale` function nested inside the first parameter, which is also nesting the `dist` function. To understand the nested functions you should start at the innermost function. If you remember the `dist` function from Chapter 3, you will know that it creates a distance matrix. Moving

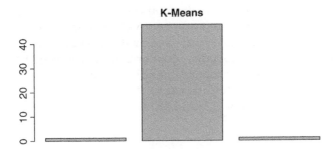

Figure 5.5 The k-means clustering with three partitions on work experiences.

outward, the distance matrix is then used within `cmdscale`. This calculates the "principal coordinates analysis" which is similar in spirit to the more common principal component analysis. Essentially `cmdscale` forces a high number of dissimilarity measures into a low-dimensional space, making it great for plotting. It is searching for the main axes of the dissimilarities matrix so either 2D or 3D space. The results of these two nested functions make up the first parameter for `plotcluster`. The second is the cluster assignments from the k-means object. The `plotcluster` function itself uses discriminant coordinates to plot the principal differences between the first and second parameters. Again this is similar to `cmdscale`, in that it is reducing high-dimensional data for a 2D plot. After running the following code you will notice that the k-means clustering approach is not very explanatory. However, this code will be useful later in the chapter when you apply other methods for comparison. Figure 5.6 reinforces the fact that k-means clustering did not partition the work experiences DTM well.

```
plotcluster(cmdscale(dist(wk.dtm)),
wk.clusters$cluster)
```

Another great way to evaluate clustering effectiveness is with a silhouette plot. To interpret a silhouette plot, think of each document grouped by their

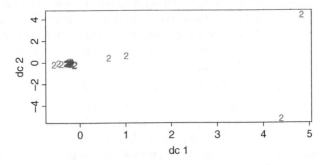

Figure 5.6 The `plotcluster` visual is overwhelmed by the second cluster and shows that partitioning was not effective.

assigned cluster casting a shadow over the visual. The more complete the shadow the more defined the cluster. A silhouette coefficient of 1 would be a perfectly defined cluster with each document creating a long shadow extending across the visual. Since the documents themselves are grouped by cluster you should expect to see distinct shadows for each partition.

To create a silhouette plot from a k-means clustering output use the following code. Once again, you calculate a distance matrix for all points in the scaled work experience DTM. Next you apply the silhouette function to the cluster assignment results. In this case, it is the assigned cluster for each document from 1, 2 or 3. Then both are simply plotted using the base plot function.

```
dissimilarity.m <- dist(wk.dtm.s)
plot(silhouette(wk.clusters$cluster, dissimilarity.m))
```

If the k-means algorithm were effective for this data you would expect to see three distinct groups of silhouettes with a large average silhouette coefficient in

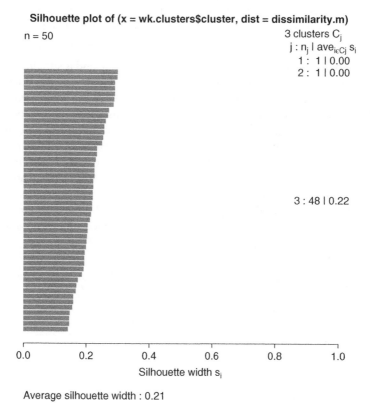

Figure 5.7 The k-means clustering silhouette plot dominated by a single cluster.

Figure 5.7. Unfortunately, the plot in Figure 5.7 shows that the algorithm did not partition well and is dominated by a single shadow. The silhouette for two clusters is actually 0.00, so this approach is not appropriate and further methods, explained next, need to be explored.

Next you can extract the prototypical words for each cluster. The poor results in the cluster and silhouette plots mean that the prototypical terms are likely not going to be informative and are also shared among the clusters.

To extract prototypical cluster terms you call `cl_prototypes` from the `clue` library. The `clue` library provides clustering ensemble methods but also has this function for extracting prototypes from each partition. To create the work cluster prototypes you apply `cl_prototypes` to the k-means object and then transpose the result in the new object shown in the code here.

```
work.clus.proto<-t(cl_prototypes(wk.clusters))
```

The resulting cluster prototype object is a matrix. Each term is a term in the work experiences and each column is the score for a particular cluster. As a result cluster has a corresponding column. A quick way to examine the cluster values visually is with a comparison cloud from the word cloud library. The code below is used to construct Figure 5.8 from scored prototype terms.

```
comparison.cloud(work.clus.proto, max.words=100)
```

The comparison cloud in Figure 5.8 is not informative and it should come as no surprise that the second cluster has no unique words. It is the dominant cluster and therefore contains words also mentioned in the other two partitions. The next section using the spherical k-means approach explores the prototypes and resulting comparison cloud with a more insightful outcome.

Overall, it may be disappointing that k-means did a poor job of partitioning the work experiences. This may be due to the fact that the work experiences are not very diverse. They represent prior work experiences from a fairly homogenous population of call center workers. Further each has shown a measure of success by lasting longer than a year. As a tenacious HR professional and text miner, you decide to try another clustering approach called spherical k-means.

Figure 5.8 The comparison clouds based on prototype scores.

5.1.2 Spherical K-Means Clustering

You can now extend your clustering knowledge to the spherical k-means approach. It is useful to try many approaches to ensure that you select an appropriate clustering method that leads to an insightful, reliable conclusion. The benefit of spherical k-means is that it handles sparse matrices very well. You may recall that the tm package's DTM and TDM are both usually very sparse, made up mostly 0 values.

The difference between k-means and spherical k-means is the way in which similarity distance is measured. In k-means, the distances are measured in Euclidean or Manhattan distance. Euclidean distance is calculated as the root sum-of-squares of differences between documents. Squaring then taking the root of distances means there are no negative distances between objects. The Manhattan distance measure is calculated as the sum of absolute distances in order to get rid of any negative distances. This Euclidean distance is the distance measured as a straight line and is used in the previous example. In contrast to Euclidean distance, spherical k-means clusters are based on using cosine similarity to calculate distance. Cosine similarity is the cosine of the angle between documents after the documents have been normalized. Similar documents will have a cosine similarity approaching 1. As documents become less similar their respective cosine similarity decreases from 1.

Figure 5.9 shows the difference in distance measurements used for both k-means and spherical k-means clustering. In the illustration document D1 has a lot to do with "text" such as an instructional book for aspiring authors. Documents D2 and D4 are a mix of "text" and "mining" like this book. Lastly document D3 is mostly about "mining" which could be mineral mining. On the left documents are plotted in 2D space with Euclidean distances shown as arrows from D4. Using Euclidean distance documents D1, D4 and D3 would be clustered since they are

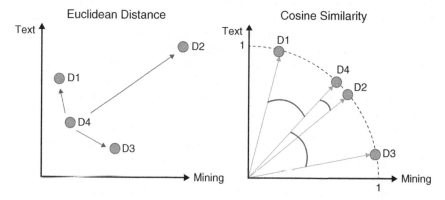

Figure 5.9 A comparison of the k-means and spherical k-means distance measures.

closest, while document D2 would be in its own cluster. On the right, the document lengths have been normalized to length 1. Once this length normalization has been performed the documents now form part of a circle leading to the spherical nature of spherical k-means. Keep in mind that since the illustration is only in two dimensions, it is not really a sphere for this example. Further on the right-hand side the distance measure is now the cosine of the angles shown as arcs from D4. In this case, the spherical k-means approach would correctly cluster documents D2 and D4 as being similar mixing "text" and "mining" more equally than D1 or D3.

You can use the following code to execute a spherical k-means approach on the example HR texts. Remember, the goal is to identify specific terms, attributes or work experiences, that can help educate HR recruiters at your call center. You hope the new approach yields more insights and a possible recommendation for your recruiting team.

Start by loading the spherical k-means and cluster ensemble libraries in addition to the previous section's libraries. To get started, the code below skips ahead past corpus creation and cleaning to recreate the work experiences DTM that has TFIDF weighting as shown in the previous k-means section.

```
library(skmeans)
library(clue)
wk.dtm<-DocumentTermMatrix(wk.corpus,
control=list(weighting= weightTfIdf))
```

Next you will actually perform the spherical k-means clustering. To do so, you invoke the skmeans function. It is applied to the work experience DTM called wk.dtm. The next parameter specifies the number of clusters similar to simple k-means. Once again, you decide to choose three clusters. The next parameter is "m." This parameter is the fuzziness of the clusters as the algorithm is constructed. This allows the cluster sphere to have a border that is not concretely defined. This is analogous to a border between two countries that have a demilitarized zone in between. Both countries could claim this fuzzy shared border. The m parameter ranges from 1 upward. A value of 1 would mean to perform a hard partition without any fuzziness between cluster borders. The higher the m value the fuzzier the borders. Here, the m parameter is 1.2, meaning some fuzziness is needed. The last section of skmeans contains a list of nruns and verbose. The nruns number tells the function to rerun the model building a specific number of times. This helps to ensure that the stability of the model results are similar to doing cross validation in other machine learning methods. The last parameter, verbose, simply tells the function to print the ongoing results to the console. Because text mining can be very computationally intensive, this helps you to know whether the model has frozen your computer, as you get updates in the console while the model is building. The spherical k-means function has multiple methods that can be executed including "genetic," "gmeans" and "kmndirs" among others. However, when doing a soft partition the only method that can be used is `pclust`. As you experiment by changing the m value you may also want to change the method within the functions too.

```
soft.part <- skmeans(wk.dtm, 3, m = 1.2, control =
list(nruns = 5, verbose = T))
```

Once again, you can create a bar plot to see how the cluster partitioning occurred. Unlike the k-means object the soft partition spherical k-means object does not have the total cluster size automatically calculated. As a result, the code below nests the table function on the individual cluster assignments before calling the bar plot.

```
barplot(table(soft.part$cluster), main='Spherical
k-means')
```

Figure 5.10 shows three clusters with more evenly distributed frequency compared to the k-means approach.

Next you create a cluster plot for visual comparison to the k-means cluster plot. Again using the `plotcluster` function you pass in a distance matrix nested in the `cmdscale` function as the first input. The second object passed in `plotcluster` represents the soft partition's 50 document assignments to the three clusters.

```
plotcluster(cmdscale(dist(wk.dtm)), soft.part$cluster)
```

Figure 5.11 is the resulting cluster plot from the spherical k-means approach. You can plainly see a lot of overlap but there is some improved dispersion among the individual documents compared to k-means. Specifically, you can see that all clusters overlap, yet some documents assigned to the second and third clusters are relatively separated. The third and first clusters overlap too, but you can also see some minor separation. As a result, some parameter tuning may be needed to further improve the results.

Next you may want to compare the silhouette plot from the previous section. The spherical k-means object contains all the information needed to calculate

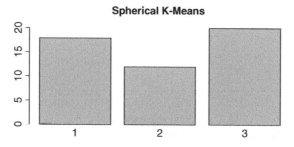

Figure 5.10 The cluster assignments for the 50 work experiences using spherical k-means.

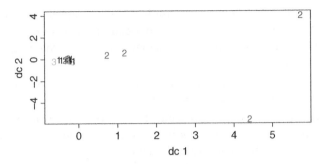

Figure 5.11 The spherical k-means cluster plot with some improved document separation.

the silhouette. Thus it is not necessary to calculate a distance plot first as was done with the k-means object. In a silhouette plot you expect to see three distinct cluster shadows within the plot since you chose three clusters to begin with. Figure 5.12 shows that three distinct clusters are emerging. This contrasts with the previous silhouette plot although the spherical k-means silhouette size is still small. Keep in mind that the corpus is small and this is merely an example but Figure 5.12 does show that it is behaving as you expect.

```
plot(silhouette(soft.part))
```

For an in-depth look at the partitions you again extract the prototype scores using `cl_prototypes`. The soft partition cluster prototype scores is a three-column matrix which can be used to create another comparison cloud. The code below is used to create the word cloud in Figure 5.13.

```
s.clus.proto<-t(cl_prototypes(soft.part))
comparison.cloud(s.clus.proto, max.words = 100)
```

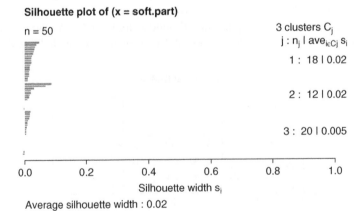

Figure 5.12 The spherical k-means silhouette plot with three distinct clusters.

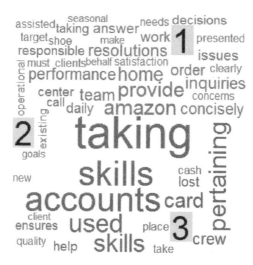

Figure 5.13 The spherical k-means comparison cloud improves on the original in the previous section.

Finally some interesting and unique words start to appear. In cluster 1, the words Amazon and home appear. In cluster 2, the most prototypical terms include responsible, daily, team, goals and performance. So you may conclude, based on this visual, that your longest tenured call center agents were at one time Amazon employees providing home resolutions. Still another cluster may be one that had an operational leadership role because they were responsible for the daily performance of their team's goals.

In order to confirm these suspicions and possibly add new insights you decide to explicitly review the top five most prototypical terms in each cluster. To do so you reference the same s.clus.proto matrix used for the comparison cloud. Each column representing a cluster is sorted in a decreasing fashion. Then you can simply review the top five terms through indexing 1 to 5. This is shown in the code snippets below for the three clusters. The top prototypical terms for each cluster are shown in Table 5.3.

```
sort(s.clus.proto[,1],decreasing=T)[1:5]
sort(s.clus.proto[,2],decreasing=T)[1:5]
sort(s.clus.proto[,3],decreasing=T)[1:5]
```

Compared to the k-means clustering approach clearly measuring the cosine similarity in a spherical k-means has shown more promising results. In practical application, you would likely be working on a larger corpus and also explore the tuning parameters when constructing the clusters. However, in this limited example, you may come to the conclusion your recruiters should focus on former

Table 5.3 Top five terms for each cluster can help provide useful insights to share.

	Cluster1	Cluster2	Cluster3
1	Amazon	Team	Taking
2	Provide	Target	Multitasking
3	Orders	Call	Accounts
4	Order	Center	Skills
5	Issues	performance	Inbound

Amazon employees, leadership and goal-focused experiences and resumes mentioning multitasking.

5.1.3 K-Mediod Clustering

Another related algorithm is the k-mediods method. K-mediods uses the median value instead of average when calculating the cluster centers. Within R you create a "pam" object. This stands for partition around median. The k-mediods does not compute "centroids" but instead the cluster centers are called "mediods." Similar to k-means approaches the k-mediods approach seeks to minimize the distance of the documents to the center point. However, instead of a random point in space used for the k-means approach, the k-mediods approach will always pick an actual document.

There are two benefits to using k-mediods with document clustering. First, since the approach utilizes actual prototypical centers rather than a calculated point in space, you will be able to get an explicit example for each cluster. The next mediod clusters are less affected by outliers. K-means clusters can be affected if there is an outlier document, meaning a document that uses terms completely unlike the rest of the collection in either diversity or frequency. K-mediod and k-mean approaches are often similar but it is worthwhile to explore many approaches during the course of an analysis.

A numeric example illustrates both benefits of k-mediods well. If you had a population with values 1, 2, 4, 6 and 100, the mean and median values differ greatly. The mean of the population is 22.6. The sum, 113, is divided by 5. Note the 22.6 is highly affected by the 100 value and that the population was made of integers, so 22.6 does not really exist in the original data. In contrast, the median value for the population is 4. This number seems more appropriate in that it is not skewed by the outlier 100 and is also equal to a value in the population. Applied to documents, this means that a mediod approach may be more insightful because the prototype exists for examination, and given the wide range of possible words, mediod clusters may be less affected by an outlier document containing highly diverse lexicons.

The following code sections execute a k-mediods clustering algorithm on the existing HR resume use case. As before, you start with the work experience DTM with TFIDF weighting. It is cleaned with the exact same method used in the previous sections for consistency.

```
wk.dtm<-DocumentTermMatrix(wk.corpus,
control=list(weighting= weightTfIdf))
```

Within the `fpc` library the function `pamk` can be used to construct a k-mediods object. As with k-means, the default distance measure is the Euclidean distance although you can specify other methods. One of the nice parameters within `pamk` is `krange`. Using a sequence of integers, `krange` allows you to specify a range of possible clusters rather than explicitly selecting them a priori. The function will calculate the silhouettes for each of the clusters in the range and return the number of clusters with the highest silhouette area.

```
wk.mediods<-pamk(wk.dtm, krange=2:4, critout = T)
```

After you run the `pamk` code you will note that two clusters have the highest silhouette and this is printed in your console. As before, you decide to create a silhouette plot using the next code snippet.

```
dissimilarity.m <- dist(wk.dtm)
plot(silhouette(wk.mediods$pamobject$clustering,
dissimilarity.m))
```

Unfortunately very similar to the k-means approach a single cluster is dominating the partitioning. One way this shows up is in the silhouette plot in Figure 5.14.

Although disappointing, for the sake of this educational example you could still create a comparison cloud. However, unlike previous examples the mediod approach means that the prototypical cluster centers are actually real documents. This can be helpful with larger and more diverse corpora to explicitly show a document that exemplifies the median center of each partition and this can help "make it real" to decision-makers. First you extract the cluster prototypes and transpose the result. Then call the word cloud library function, `comparison.cloud` on the prototype object. Here the mediod points are work experiences 15 and 40 and are labeled explicitly. If you had more clusters and therefore mediod centers you would have more than two documents in the comparison cloud in Figure 5.15.

5.1.4 Evaluating the Cluster Approaches

Revisiting the six-step text mining process, you should now feel comfortable reaching an insight or recommendation. Reviewing the original intent of the exercise below you can safely come up with some recommendations for your call center recruiters to focus on.

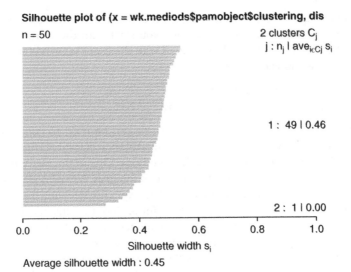

Silhouette plot of (x = wk.mediods$pamobject$clustering, dis

n = 50

2 clusters C_j

$j : n_j \mid ave_{i \in C_j} \, s_i$

1 : 49 | 0.46

2 : 1 | 0.00

0.0   0.2   0.4   0.6   0.8   1.0

Silhouette width s_i

Average silhouette width : 0.45

Figure 5.14 K-mediod cluster silhouette showing a single cluster with 49 of the documents.

7) **Reach an insight or recommendation.** The goal is to identify three distinct terms in employee clusters that may help recruiters identify candidates likely to stay more than a year, based on past experiences.

The best clustering approach in this limited exercise was spherical k-means. You could likely improve the results by using a larger corpus and performing clustering on contrasting documents, e.g. work experiences less than a year. It may not always be the case but here the spherical k-means partitions appear to be more valid and some conclusions can be drawn. The prototypical terms in each cluster can help lead you to the conclusion that call center agents that last more than a year have the following attributes in their prior experiences:

- performance target oriented
- worked at amazon.com providing order resolution
- list multitasking and inbound [calls] in prior experience

For a statistical comparison of cluster outputs, you may want to use the function `cluster.stats` from the `fpc` package. This function will calculate relevant evaluation statistics, allowing cluster approaches to be compared. This can help to validate the best approach and the number of clusters and help in decision-making. The code below applies this function to the k-means and k-mediod approaches. In the code chunk, the first parameter is a distance matrix based on the DTM, followed by the cluster assignments from each of the two algorithms.

```
results<-
cluster.stats(dist(wk.dtm.s),wk.clusters$cluster,
wk.mediods$pamobject$clustering)
```

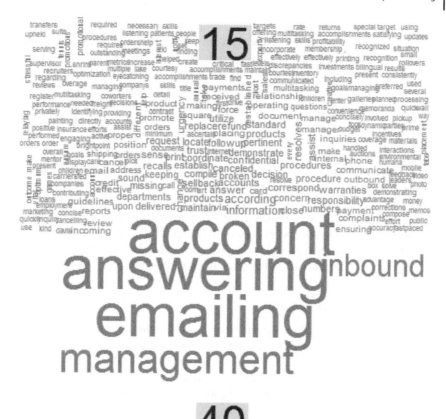

Figure 5.15 Mediod prototype work experiences 15 & 40 as a comparison cloud.

The resulting object is a list with 34 pieces of information. Overall these statistics are beyond the scope of this basic text mining book. However, items like average silhouette width should be relatively familiar, since you have already made silhouette plots in this chapter. Other items include the average distance of documents within a cluster to its center, and average distances of a point in the cluster to the points of other clusters. If you want to explore these statistics in greater detail, a book on unsupervised clustering techniques will go into much greater detail.

5.2 Calculating and Exploring String Distance

So far in this chapter we have followed the six-step text mining process for document clustering in an HR analytics case study. In this next section we use a different type of distance measurement. We will work to find the distances between strings in a different manner than Euclidean distance or cosine similarity. This section introduces Mark van der Loo's `stringdist` package and another corpus to work on. For this section we will not work within the structured text mining workflow so we can focus on the concepts. However, in the topic modeling section we will work on another case study.

The string distance package has three important functions within the context of text mining (it can also work on integers). The first `amatch` (and related `ain`) compute a fuzzy match between a character string and a table of other strings. The result of the function is the position in the table that is a fuzzy match. The `stringdist` function calculates the distance between a string and a vector of various strings. Still a third useful function in this package is `stringdistmatrix` used to calculate a distance matrix that can be used in clustering. Both `stringdist` and `stringdistmatrix` return a number representing the number of basic operations needed to turn one string into the other. This contrasts with `amatch`, which returns the position of a fuzzy match within a table. Do not be overwhelmed, string distance and the accompanying functions are explained here.

5.2.1   What is String Distance?

String distance is the measurement of one string to another at the character level. That is to say the number of letters that need to be changed in some manner to make that string into another. For example, the word "cat" has a distance of 1 to the word "bat." This is because the substitution of "c" for "b" is the only difference. The distance between "cat" and "bats" is two because there is a single substitution and also an insertion of "s."

All three functions, `amatch`, `stringdist` and `stringdistmatrix`, are passed a specific method allowing for the types of changes that are allowed. The five methods dictate which of the four operators can be used to change one string to another so they match. These include substitution, deletion, insertion and transposition of adjacent characters. The four operators work within the larger character string and act upon substrings. The distance is then just the minimum number of changes needed to match the two terms. Another practical example is "book" and "books." These strings have a distance of 1. This is because you can insert a single "s" to get a match between the two terms.

The five distinct methods passed to the functions represent a mixture of one or more of the four substring operators. Fuzzy matching and string distance measurements can be the basis behind document spellcheck and corrected search queries when using a search engine.

The first of the five methods uses only substitution. As you may expect, this is when you can substitute one character for another in order to match the terms. For example, assume that you only had substitution as your tool to match two terms. When comparing "racecarz" and "racecars" you would be able to substitute one character to make them match. The single letter "z" can be substituted with "s" to get a match. If a term was further misspelled to "racearzc" the substitution count increases. In this case, there are four substitutions, the second "a" to "c", second "r" to "a", "z" to "r" and the last "c" to "s" to make a match. Notice that the substitution operations did not insert a "c" and move the "ar" to the right. It simply substituted character by character. This distance method is called "Hamming."

Still another method that can be applied to string distance functions is optimal string alignment (OSA). The OSA method allows the use of all four operations, substitution, deletion, insertion and transposition. Armed with more tools, our racecar example results change. With OSA, racecarz and racecars still have distance of 1 stemming from the substitution operator. However, given that OSA can use more operations to match strings, the example of "racearzc" and "racecars" has a distance of 3 not 4. Since OSA can use insertion it inserts a "c" in front of the "ar" as the first operation. Next, OSA can delete the last letter and substitute the "z" to an "s". So the total number of operations in this example is only three. It should be noted that in OSA the transposition limit is one. This differentiates it from other distance methods and means that a character cannot be transposed twice to move it elsewhere in the larger string.

Related to the OSA method is the Damerau–Levenshtein (DL) method. With this methodology, all operations are available, just like OSA. The difference between OSA and DL is that adjacent substrings can be transposed more than once.

The last two methods are less common and more restrictive in the types of operators used. The Levenshtein distance method uses substitution, deletion and insertion only. It is not permitted to use transposition. The longest common string (LCS) method only uses deletion and insertion. LCS is not permitted to change the order of characters and returns the total characters not shared in a substring. For example, the distance between "apple" and "crabapple" is four using LCS. This is because "apple" is the longest shared substring between the two terms. That leaves "crab", which is four characters long. However, for "crabapples" and "apple" the longest common string calculation would now be 5. This is because of the original four letters of "crab" and the additional "s."

The best way to get fluent with the string distance measures is to use them. Simply change the method part of the code below as well as the input strings to get the change in scores.

```
> stringdist('crabapple','apple',method="lcs")
[1] 4
> stringdist('crabapples','apple',method="lcs")
[1] 5
```

The five methods of strings distance calculation are summarized in Table 5.4 along with the parameter abbreviation used in the `stringdist` library functions.

Table 5.4 The five string distance measurement methods.

Method	Substitution	Deletion	Insertion	Transposition	Function abbreviation
Hamming	Yes	No	No	No	`hamming`
OSA	Yes	Yes	Yes	Yes, Only Once	`osa`
Damerau–Levenshtein	Yes	Yes	Yes	Yes	`dl`
Levenshtein	Yes	Yes	Yes	No	`lv`
Longest common string	No	Yes	Yes	No	`lcs`

There are other distance measures in the library that can be explored, although these five are common. The other measures include soundex for phonetic similarities and Jaro, which measures between 0 and 1. When performing string distance analysis it is important to know the method with which the strings were collected, the ways distance calculations impact results and subsequently the resulting cluster analysis. Damerau–Levenshtein is popular because it has all available operators, and the results are usually stable. LCS can produce wide ranges of distances which may be misleading to interpret but useful for clustering. If the strings were transcribed or dictated, then the soundex method may also be useful. In the end it is the judgment of the text miner to use a method that is both accurate and effective. For the `stringdist` library, the default method is OSA.

5.2.2 Fuzzy Matching – Amatch, Ain

The `amatch` function is fairly straightforward and similar to the base R function `match`, the latter returning the position of the first exact match between two terms. The `amatch` function, however, provides a bit more flexibility. It returns the position of the closest term or "fuzzy" match between strings given a maximum distance allowed. It receives a string that you want to approximately match against a group of words along with the distance number and method. Consider the following code lines using `match` and `amatch`.

```
match('apple',c('crabapple','pear')
[1] NA
```

Here `match` returns NA because there is no exact match between apple and the other two terms.

```
amatch('apple',c('crabapple','pear'),maxDist=3,
method='dl')
[1] NA
```

In this example, the result is the same, NA, despite having all operators to work with because "dl" is the method specified. This is because there is no approximate matches that have a maximum distance of 3 between apple and the other terms. Remember that distance is the minimum number of operators, substitutions, deletions, insertions and transpositions needed to match the terms.

```
amatch('apple',c('crabapple','pear'),maxDist=4,
method='dl')
[1] 1
```

In this case, the maxDist parameter was increased to 4. The strings "apple" and "crabapple" are separated by four deletion operations. The function returns a 1 because "crabapple" is the first term in the vector of "crabapple" and "pear."

Instead of the position of the first approximate match, you may need a True or False outcome. The ain function will provide this binary outcome and follows a similar functional pattern.

```
ain('raspberry',c('berry','pear'),maxDist=4,
method='dl')
[1] TRUE
```

In this example, four characters can be deleted so that "raspberry" and "berry" can match. The maximum distance is 4 and the Damerau–Levenshtein method is allowed to use the deletion operator. The net result is True. Changing the method to method= 'hamming' will result in a False being returned. This is because the Hamming method is not allowed to use deletion. Similarly, had the "berry" term been "Berry" instead the ain function would return False. This is because the function is case sensitive. To normalize the strings you would apply tolower to the terms being passed to the function. Hopefully you are familiar with tolower, because it is a function used in cleaning and preprocessing earlier in the book.

5.2.3 Similarity Distances – Stringdist, Stringdistmatrix

The next functions stringdist and stringdistmatrix do not return the position of an approximate match from a vector but instead return the actual minimum number of operators needed to make a match between terms. Stringdist works on a single term and one or more terms while stringdistmatrix will perform the calculation between all strings.

The stringdist function calculates the distance between a single term and another single term or vector containing multiple terms. In this example, the distance of the single term "raspberry" is being calculated between "berry" and "pear" using the Hamming method.

```
stringdist('raspberry',c('berry','pear'),method='hamm
ing')
[1] Inf Inf
```

The result is infinity twice. This is because there is no way to match raspberry to either term using only substitution. However, changing the Hamming to the more robust optimal string alignment in the following code will return two integers. Since OSA is the default, you do not really need to specify it as shown here.

```
stringdist('raspberry',c('berry','pear'),
method='osa')
[1]  4  6
```

The first integer returned is 4. This is because the default method OSA can delete four characters to match "raspberry" and "berry". The next distance between "raspberry" and "pear" is a bit more complicated. The first three letters, "ras", are all deleted. That leaves a substring "pberry". Next delete the "b" and "y". The resulting string is "perr" and there is still one more operator needed to match "pear". The function then substitutes the first "r" for an "a" to make the match. Figure 5.16 may help illustrate the operations needed to match "raspberry" to "pear".

Earlier in the chapter you worked on various distance measures of text to create distance matrices and ultimately clustering. String distance is just another method for performing distance calculations and as a result can likewise be used for clustering. Now that you have a foundation in string distance calculation methods you can work on clustering. In this basic example, the code below creates a distance matrix using some of the fruit examples shown before and then plots a dendrogram.

```
fruit<-c('crabapple','apple','raspberry')
fruit.dist<- stringdistmatrix(fruit)
```

The `fruit.dist` object is a distance matrix between all combinations of the fruit names. The results are shown in Table 5.5.

When plotting a simple dendrogram illustrating the hierarchical nature of the text you should expect that "apple" and "crabapple" should be a cluster

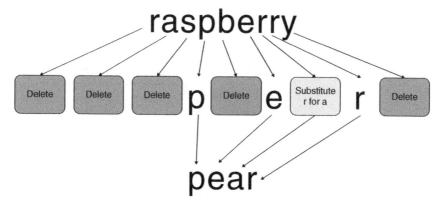

Figure 5.16 The six OSA operators needed for the distance measure between "raspberry" and "pear".

Table 5.5 The distance matrix from the three fruits.

	Crabapple	Apple
Apple	4	
Raspberry	8	6

since their distance is 4. "Raspberry" will be by itself as an outlier because it is very far from "crabapple." Since a dendrogram reduces information, the smaller difference between "apple" and "raspberry" is lost. The code below creates the expected dendrogram shown in Figure 5.17.

```
plot(hclust(fruit.dist),labels=fruit)
```

You can now apply this technique to larger strings, but beware that the dendrograms can become dense, and additional steps may be needed to make the outcomes useful. However, string distance can be useful for term aggregation and spellcheck. The examples here are meant to lay a foundation for further exploration to unlock the underlying structure and connections within a corpus using string distance.

5.3 LDA Topic Modeling Explained

Topic modeling is a probability-based approach to finding clusters within documents. It is unsupervised because you do not have document-assigned classes like "spam" versus "not spam." Instead, topic modeling observes the word

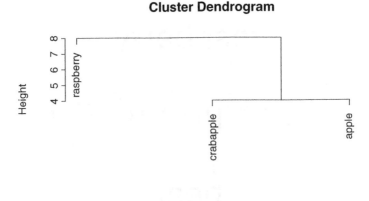

Figure 5.17 The example fruit dendrogram.

frequency distribution among all documents to define "K" topics, and documents are given a probability for each topic. As the text miner, you input the number of topics as the "K" parameter. The topic modeling algorithm assigns a probability from all observed topics to each document. With K = 4, each document would get four probabilities for being part of the observed topics.

The LDA stands for Latent Dirichlet allocation. It is considered latent because the modeling technique identifies concealed topics that are not explicitly defined by the text miner. It seeks to find the concealed group of words that represent each topic or theme. Dirichlet distributions are used in statistics to understand multivariate, or in this case multi-word, probability distributions. Dirichlet distributions can be used to model seemingly random occurrences such as rolling dice. In fact, dice rolling is not completely random because of manufacturing processes and physics. Taken holistically one may think that word choice in a large corpus is close to random, but the Dirichlet distribution can be used to model the word allocations.

The LDA topic modeling technique is not that complicated but it can be intimidating. In practice, you may not really need to possess a deep understanding to gain meaningful insights. A simple example could be using LDA topic modeling on multiple art history books. In the large corpus, books could be about painting and photography or just a single style of painting or exclusively about photography. Topic modeling may observe the words in the books and define K = 3 topics which could include "art," "photo" and "painting." So each book is assigned a probability of being part of each of these three topics. One book may be about photography and score highly in that topic while another may be about impressionist painting and likewise have a high probability for the "paint" topic. The photography book would likely have a low probability for the "paint" topic and similarly the impressionist book would have a low score for the "photo" topic. However, all books would have a similar probability for the "art" topic. The books themselves are a mixture of topics simultaneously.

The topic modeling approach seeks to answer two basic questions. First, how to decide if a specific word is part of a particular topic among other topics. Second, how common is a particular topic within a document.

If the LDA modeling approach were a person she would approach a document armed with multiple differently colored highlighters. Each highlighter would be represent a topic and be defined before she started reading. As the algorithm "reads," she would highlight a word based on the frequency and code it by color to a topic. If applied to this book the algorithm would highlight words having to do with case studies in "blue," math-related terms in "yellow" and R code in "red." Once this is completed the words in each color group are tallied. The list of blue highlighted words represents one particular topic and so on. Within the book, each chapter could be viewed as a mixture of these topics. Some chapters may have more math, which would be yellow, while other chapters have more case-study-related words which would be blue. Figure 5.18 illustrates this technique on an earlier passage in the chapter. The passage is a mix of all three topics but mostly math terms.

In this section, we examine a case study and a particular type of LDA topic modeling called Gibbs sampler. This type of LDA model is faster than others and can be useful when working on large corpora. The Gibbs sampler is a Markov chain Monte Carlo (MCMC) algorithm. While complicated, it approximates distributions, then correlates Markov chain samples and is used for statistical inference techniques. In fact, the topic modeling algorithm is doing just that, sampling to infer distributions of words which is in turn inferring the topics within documents. Beware that sampling can decrease accuracy. However, increasing the number of sampling iterations in the model will help to ensure stable results. There are other topic modeling R libraries that you may want to explore beyond this case study including `topicmodels`, `irlba` and `textmineR`.

5.3.1 Topic Modeling Case Study

In this section, we revisit the six-step text mining process in a new case study. Using articles from the Guardian newspaper as our corpus, you will do some topic modeling to create an interactive visual of the topic model and a treemap representing polarity, topic and size of article.

1) **Problem definition and specific goals.** The goal of this project is to analyze Guardian newspaper articles to understand the topics covered, the journalistic effort put towards each topic and the polarity of the articles. The specific problem statement could be "How does the Guardian newspaper prioritize articles about Pakistan?"

2) **Identify the text that needs to be collected.** This corpus was collected from the GuardianR package, using the newspaper's API. This API will be discussed in detail later in the book but the topic modeling section has already collected the articles. The intended text contains all Guardian articles mentioning Pakistan between November 14 2015 and December 1 2015.

Getting the Data

Please navigate to www.tedkwartler.com and follow the download link. For this analysis, please download "Guardian_articles_11_14_2015_12_1_2015.csv." This file contains 27 articles and associated information from the API.

---

3) **Text Organization.** The articles will be organized in the returned API object chronologically. The text will be cleaned using common and new character manipulation functions.

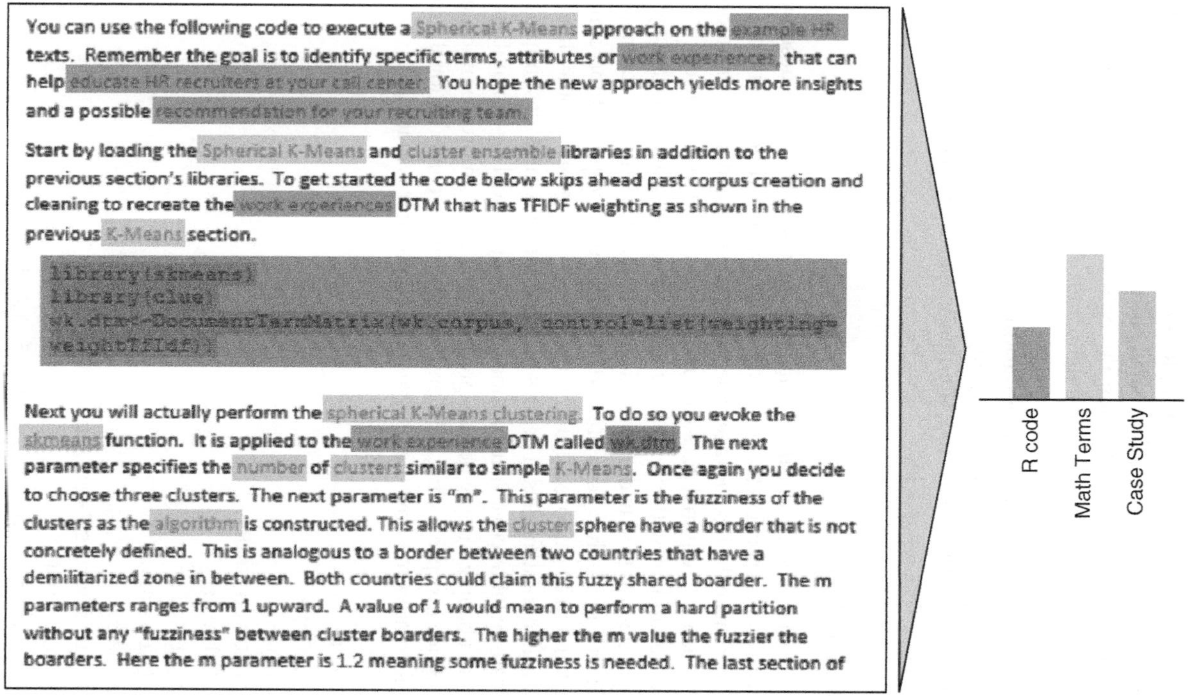

You can use the following code to execute a Spherical K-Means approach on the example HR texts. Remember the goal is to identify specific terms, attributes or work experiences, that can help educate HR recruiters at your ad centre. You hope the new approach yields more insights and a possible recommendation for your recruiting team.

Start by loading the Spherical K-Means and cluster ensemble libraries in addition to the previous section's libraries. To get started the code below skips ahead past corpus creation and cleaning to recreate the work experiences DTM that has TFIDF weighting as shown in the previous K-Means section.

```
library(skmeans)
library(clue)
WK.dtm<-DocumentTermMatrix(wk.corpus, control=list(weighting=
weightTfIdf))
```

Next you will actually perform the spherical K-Means clustering. To do so you evoke the skmeans function. It is applied to the work experience DTM called wk.dtm. The next parameter specifies the number of clusters similar to simple K-Means. Once again you decide to choose three clusters. The next parameter is "m". This parameter is the fuzziness of the clusters as the algorithm is constructed. This allows the cluster sphere have a border that is not concretely defined. This is analogous to a border between two countries that have a demilitarized zone in between. Both countries could claim this fuzzy shared boarder. The m parameters ranges from 1 upward. A value of 1 would mean to perform a hard partition without any "fuzziness" between cluster boarders. The higher the m value the fuzzier the boarders. Here the m parameter is 1.2 meaning some fuzziness is needed. The last section of

Figure 5.18 "Highlighters" used to capture three topics in a passage.

4) **Feature Extraction.** The lda package provides a new function called lexalize which captures document level information. This is in contrast to the DTM or TDM objects we have created thus far.

5) **Analysis.** First you will create a topic model, reviewing the words within each topic. Then plot the model visually in a dynamic plot that is opened in a web browser. Next, you will create a treeplot, illustrating three dimensions including polarity, length of document and observed topic from a topic model prediction.

6) **Reach an insight or recommendation.** Using the topic modeling and visualizations you will be able to group the articles by observed topic and then hopefully understand how the newspaper's writers spend their time concerning Pakistani articles.

5.3.2 LDA and LDAvis

This script is slightly different from our previous ones. This is because we are going to clean and preprocess the text vector of 27 articles without a custom function. This is because the lda package expects a text vector and not a corpus object.

To begin, load the required packages and the text object from the csv. By now you should be familiar with the tm and qdap packages. The next package, lda, provides functions to read in documents and fit latent Dirichlet allocation (LDA) models. Once a model has been created, it can be used to predict topic models on new text and to be explored. Later in the chapter we review another lda package simply called topicmodels. The next package, LDAvis, provides simple functions to create an interactive visualization of the topic model for exploration. The last package, pbapply, simply adds a progress bar to the more common apply functions. Since text processing on many articles can be computationally intensive it is informative to know the progress your computer is making.

```
library(tm)
library(qdap)
library(lda)
library(GuardianR)
library(pbapply)
library(LDAvis)
library(treemap)
library(car)
options(stringsAsFactors = F)
text<-read.csv(
'Guardian_articles_11_14_2015_12_1_2015.csv')
```

Step 3. Text organization – The text object has many extraneous columns. You are only going to analyze the text$body vector which contains the text for each of the articles. The following code block cleans that single vector in a new object called `articles`. First `iconv` is used to convert the character vector, `text$body`, from latin1 to ASCII while simultaneously creating the `articles` object. This is not always needed but sometimes API text has unusual encodings that need to be changed. Next the `gsub` function is used to remove any url links that start with http. The raw API text contains many urls that can hinder the topic modeling by emphasizing the urls instead of the actual article text. Next the `bracketX` function from qdap is applied. This helps to remove all text within brackets. For example, text within square brackets is often added to quotes as in "[inserted text]." In this case, `bracketX` is applied to the articles because the API returned html text such as "
." The "br" line break in html is not really part of the article and thus needs to be extracted. Next, the more familiar `removeNumbers`, `tolower` and `removeWords` are applied to the articles.

```
articles <- iconv(text$body, "latin1", "ASCII", sub="")
articles <- gsub('http\\S+\\s*', '', articles)
articles <- bracketX(articles,bracket='all')
articles <- gsub("[[:punct:]]", "",articles)
articles <- removeNumbers(articles)
articles <- tolower(articles)
articles <-removeWords(articles,c(stopwords('en'),
'pakistan','gmt','england'))
```

The lda package struggles with blank word tokens made entirely of spaces. It will treat empty tokens made of spaces as actual words, which will seem very frequent to the model. If they are present, this skews the results of your LDA model. Also, this has a downstream impact on the LDAvis interactive plot. The interactive visualization enforces that all word tokens in the articles' vocabulary have at least one character.

There are many solutions to trimming the words so that they have at least one character, but the code below creates a simple function called `blank.removal` using string splitting. You can also adjust the zero in the subset line to focus the analysis on longer and possibly more insightful words. As a result the function becomes another tuning parameter of your analysis, giving you flexibility to extract important insights.

```
blank.removal<-function(x){
  x<-unlist(strsplit(x,' '))
  x<-subset(x,nchar(x)>0)
  x<-paste(x,collapse=' ')
}

articles<-pblapply(articles,blank.removal)
```

To understand this function, let's explore a simple example. Consider the following two sentences "Text mining is a   good time" and "Text mining is a good time". Both sentences are the same with the exception of the extra spaces in the first sentence. The code snippet creates the example text to explore.

```
ex.text<-c('Text mining is a      good time',
'Text mining is a good time')
```

When applying `strsplit` on the first sentence the returned list has many empty tokens.

```
strsplit(ex.text[1],' ')
[[1]]
 [1] "Text"    "mining" "is"      "a"        " "
 [6] " "       " "      " "       " "        " "
[11] " "       "good"   "time"
```

In contrast, applying the string split using spaces on the second sentence shows no empty tokens.

```
strsplit(ex.text[2],' ')
[[1]]
[1]  "Text"   "mining" "is"     "a"       "good"   "time"
```

In the custom function, the string split is nested in unlist, which simply changes the list to a character vector of each individual word including the empty ones. The code below creates a simple object, and demonstrates the character vector outcome.

```
char.vec<-unlist(strsplit(ex.text[1],' '))
char.vec
 [1]  "Text"   "mining" "is"     "a"       " "       " "
 [7]  " "      " "      " "      " "       " "       "good"
[13]  "time"
```

The next function subsets that vector of words by number of characters, nchar, greater than 0. Then the console results are printed showing the removal of the empty words.

```
char.vec<-subset(char.vec,nchar(char.vec)>0)
char.vec
```

```
[1]  "Text"    "mining" "is"      "a"      "good"    "time"
```

At this point, the sentence is still broken up into individual words, but now with at least one character. The last part of the `blank.removal` function merely collapses the text back into a single sentence or document.

```
char.vec<-paste(char.vec,collapse=' ')
char.vec
[1]  "Text mining is a good time"
```

So applying the custom function `blank.removal` not only solves the empty token problem but also allows you to have a tuning parameter during analysis. As a result, the function is better than simply trimming tokens using regular expressions.

The `blank.removal` function is applied to the document vector using the `pblapply` function. You could also use `lapply` from base R but the pb version adds a progress bar so you can see how long it will take. This can be helpful if you are working on a lot of text.

```
articles<-pblapply(articles,blank.removal)
```

Step 4. Feature extraction – Step 3, text organization, is complete. The text has been preprocessed and cleaned within the document vector. Now you are going to create your feature extractions in the fourth step of the workflow. The lda package does not use the more familiar volatile corpus from `tm` or word frequency matrix from qdap. Instead the lda package provides a function `lexicalize`. The list created from `lexicalize` contains two elements called "documents" and "vocab." The result is fairly complicated, so we will first look at a simple example before applying it to the Guardian articles.

Consider the following sentence in the `ex.text` object below that is lexicalized.

```
ex.text<-c('this is a text document')
ex.text.lex<-lexicalize(ex.text)
```

The resulting list has two elements called documents and vocab that we will examine. The `ex.text.lex$documents` is a nested list of matrices. Each document has its own matrix of word frequency similar to a DTM. Examining our corpus of a single document is shown next. You will see that the document is represented with four columns, one for each word. The matrix also has two rows. The first represents the position of the word in the vocab and the second is the frequency of that word in the document. With this information and the vocab object you can reconstruct the original document. In this illustration the document has one for each word in the vocab starting with token zero (not one).

```
ex.text.lex$documents[[1]]
     [,1] [,2] [,3] [,4]
[1,]    0    1    2    3
[2,]    1    1    1    1
```

Reviewing the vocabulary in the ex.text.lex object returns all four of the unique word tokens from the corpus. The vocabulary is a text vector of unique words.

```
ex.text.lex$vocab
[1] "this"      "is"          "a"           "document
```

The quixotical nature of the lexicalize object is increased with more non-unique terms and additional documents. In this next example, another document is added to the first and then lexalicalized.

```
ex.text<-c('this is a document', 'text mining a text
document is great')
ex.text.lex<-lexicalize(ex.text)
```

The resulting ex.text.lex$document object now has two matrices shown below and referenced by double brackets. The first is the same as before. The second matrix has 4,5,2,4,3,1,6 in the first row, referencing the position of the word in the vocabulary. Notice that the frequency of the words is always one in row 2 in this matrix, despite the fact that "text" appears twice. This contrasts the lexicalize matrices from DTMs shown before.

```
[[1]]
     [,1] [,2] [,3] [,4]
[1,]    0    1    2    3
[2,]    1    1    1    1
[[2]]
     [,1] [,2] [,3] [,4] [,5] [,6] [,7]
[1,]    4    5    2    4    3    1    6
[2,]    1    1    1    1    1    1    1
```

When reviewing the vocabulary you can recreate the original documents based on the specific matrix. Remember that the first vocabulary word is referenced as 0 not 1.

```
ex.text.lex$vocab
[1] "this"  "is"    "a"     "document" "text"   "mining"
"great"
```

So the second document vocab positions are 4,5,2,4,3,1,6 indexing to "text," "mining," "a," "text," "document," "is" and "great."

Now that you have a basic understanding of the `lexicalize` results, you can apply it to your Guardian articles.

```
documents <- lexicalize(articles)
```

Step 5. Analysis – At this point, you have extracted the information from the text and can begin step 5, analysis. You will need to calculate some summary statistics to be used later. The lda function `word.counts` computes the word count for the set of documents. The result is similar to a word frequency matrix where the total number of times a word is used throughout all documents is computed in a table. The function accepts the documents and the explicit vocabulary to perform the tabulation. The other function `documents.length` is a summed vector for the total number of words in each document. So, long articles containing many words will have larger numbers. Using the previous example the document length is a sum of the `ex.text.lex$document` row containing ones.

```
wc <- word.counts(documents$documents,
documents$vocab)
doc.length<- document.lengths(documents$documents)
```

Further analysis beyond these two summary objects is needed. More importantly you are going to fit the LDA model. To begin, specify function parameters k, iterations, alpha and eta. The k variable represents the number of topics the function will identify. For example, if you specify k = 10, the function will observe 10 topics from the corpus. It is good to revise this number to identify an appropriate number of distinct topics. Then iterations is the number of sampling repetitions to be done on the text. This helps ensure that you have a reliable model similar to cross validation. Next, alpha represents the prior document-topic distributions, and eta is the parameter setting the prior topic-term distributions. The code below separates the input parameters and model building, as well as setting a specific seed for reproducibility.

```
k <- 4
num.iter <- 25
alpha <- 0.02
eta <- 0.02
set.seed(1234)
fit <- lda.collapsed.gibbs.sampler(documents =
documents$documents, K = k, vocab = documents$vocab,
num.iterations = num.iter, alpha = alpha, eta = eta,
initial = NULL, burnin = 0,compute.log.likelihood =
TRUE)
```

The other inputs to the model were the `documents$documents` matrices and `documents$vocab` that were previously created. Here the model was built with 25 iterations, and eta and alpha of 0.02. The model was fit with k = 4 to extract four topics. After fitting the model, a visual inspection of `fit$log.likelihood` illustrates that the algorithm has reached a stable convergence. This is accomplished with the code below and illustrated in Figure 5.19.

```
plot(fit$log.likelihoods[1,])
```

Once the model object has been created, the lda package provides some easy functions to extract the topic words and identify the prototypical documents for each topic. Using `top.topic.words` on the `fit$topics` object along with the number of words you want to review in the code below will print the top words for each of the k (4) topics.

```
top.topic.words(fit$topics, 7, by.score=TRUE)
        [,1]          [,2]         [,3]       [,4]
[1,]  "new"         "said"       "ball"     "three"
[2,]  "australia"   "paris"      "pm"       "world"
[3,]  "test"        "attacks"    "th"       "women"
[4,]  "smith"       "people"     "four"     "far"
[5,]  "zealand"     "french"     "leg"      "attack"
[6,]  "day"         "isis"       "single"   "change"
[7,]  "voges"       "syria"      "wicket"   "will"
```

Here you can see that the first topic contains terms such as "new," "australia," "test," "smith" and "zealand." The third topic mentions "ball," "single" and "wicket." Both of these topics are about test cricket matches with Australia and New Zealand. The second topic mentions words like "isis," "paris" and "attacks." Articles with this assigned topic are about Parisian attacks. The last topic is a bit more ambiguous concerning some "attack." We could rerun the model, changing the number of topics to three, or more than four, in an effort to gain some clarity.

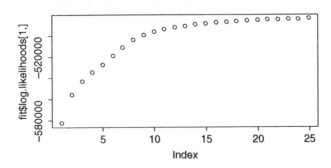

Figure 5.19 The log likelihoods from the 25 sample iterations, showing it improving and then leveling off.

The next function is used to identify the document or article that best represents each topic. Within the function top.topic.documents you pass the document sums from the model and the top number of returned articles. In this case, you are passing one so you get one article for each of the four topics. The result is a matrix showing that the 49th document is the best example of the first topic, the 50th is associated with the second and so on.

```
top.topic.documents(fit$document_sums,1)
[1] 49 50  6  3
```

To create the interactive plot using LDAvis, you need to estimate the document-topic distributions. This is done in the code snippet here, to create the theta object and referencing alpha. You also need to estimate the topic term distributions, phi, using eta.

```
theta <- t(pbapply(fit$document_sums + alpha, 2,
function(x) x/sum(x)))
phi <- t(pbapply(t(fit$topics) + eta, 2, function(x)
x/sum(x)))
```

Now that all model and article statistics are calculated you are ready to construct the interactive visual. To do so use the createJSON function and pass in the objects that were created previously. This constructs a a large character string organized as a JavaScript object notation (JSON). The JSON object is passed to the web browser in the final function.

```
article.json <- createJSON(phi = phi,theta = theta,
                   doc.length = doc.length, vocab =
documents$vocab,
                   term.frequency = as.vector(wc))
```

This last function will create all JavaScript and HTML files so it can be opened in a browser. If you are using R Studio click the "show in new window" icon to open a browser after executing this function. Figure 5.20 is a screenshot of the resulting visualization. Since it is running live you will not be able to type any other R code into your console until you explicitly stop the function. serVis(article.json)

In Figure 5.20 the left-hand side is a principal component analysis among the words in the topics. On the right is a bar chart of frequent words among all articles and also a comparison for a specific topic once it is highlighted. Shown here, topic 1 is highlighted. As expected, topics 1 and 3, related to cricket, have an overlapping topic. As you navigate the visual you will be able to understand how top words are more strongly aligned with specific topics, and how separated the topics are.

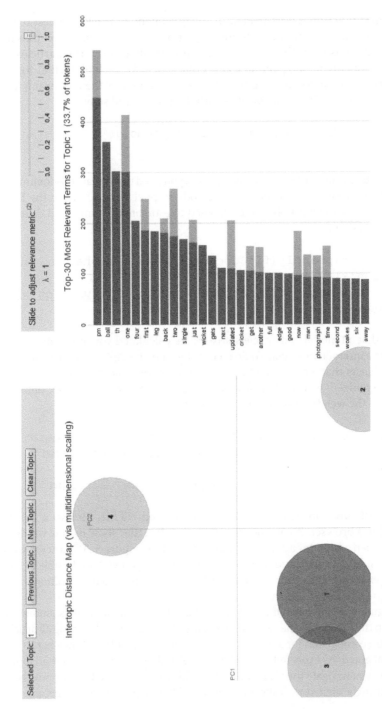

Figure 5.20 A screenshot portion for the resulting topic model visual.

The interactive visual should be informative as you explore the topics in your web browser but now we turn to constructing a treemap. A treemap illustrates hierarchical data structures. The treemap will show articles grouped by topic, the size of each article measured in number of words will be represented as the visual area and the color of each section will be the basic emotional polarity.

Since articles are a mixture of the topics you need to assign each article to the most prominent topic from the four. To do so you can create a custom function. The lda model output contains an object called assignments. This is a vector of numbers, one per word for the entire article. The number for each word corresponds to each topic. So tabulating the most frequent number among zeros, ones, twos and threes will identify an article as primarily made of that topic. In essence, you are calculating the mode for each of the document topics. One way to do this is in the function doc.assignment below.

```
doc.assignment<-function(x){
  x<-table(x)
  x<-as.matrix(x)
  x<-t(x)
  x<-max.col(x)
}
```

Each document's word topic assignments is passed to the function. The vector of numbers, one per word, is tabulated into a table, then converted from a table to a matrix. Next it is transposed, and the maximum column is returned. For example, the code below will print the topic assignments for the first 10 words from the second article. This shows that the first word is assigned to topic 2 and so on.

```
fit$assignments[[2]][1:10]
 [1] 2 1 3 3 2 1 3 3 1 0
```

Next, the custom function creates a table of the assignments.

```
table(fit$assignments[[2]][1:10])

    0 1 2 3
[1] 1 3 2 4
```

The next functions simply change the table to a matrix, and transpose it. This keeps it in the same shape as the table so you can then apply max.col.

```
t(as.matrix(table(fit$assignments[[2]][1:10])))
     0 1 2 3
[1,] 1 3 2 4
```

Using `max.col` on this matrix will return the column name containing the most frequent topic assignment. The fourth topic is the most frequent in this small sample, despite the words being diverse among the four topics.

```
max.col(t(as.matrix(table(fit$assignments[[2]]
[1:10])))) 
[1] 4
```

Now that you understand the function, it is applied to all words within each individual document or article. Using the progress bar version of `lapply` the function is applied to the model's assignment object. The entire procedure is nested in `unlist`, so the topic assignments are collapsed to a vector. The net result is a vector containing one number for each article that corresponds to the most frequent topic.

```
assignments<-
unlist(pblapply(fit$assignments,doc.assignment))
```

You can recode the topics to something more familiar using `recode` and the code below. After examining the interactive visual you may decide to change the topic number to a specific name. You could simply use the numbers or pass the top words per topic as examined previously. Here the topics are named, "Cricket1," "Paris Attacks," "Cricket2" and "Unknown." Note each recoded value is separated with a semicolon not a comma as is expected in other R functions.

```
assignments<-recode(assignments, "1='Cricket1'; 
2='Paris Attacks'; 3='Cricket2';4='Unknown'")
```

Next create a sequence of numbers corresponding to the articles in the text. Then calculate the polarity of all articles using the polarity function from qdap. For this example, the subjectivity lexicons are not adjusted. Depending on the system RAM, this may take longer than previously because articles can be relatively long. The code below only captures the polarity scores from the overall polarity object which includes extraneous information for the treemap.

```
article.ref<-seq(1:nrow(text))
article.pol<-polarity(articles)[[1]][3]

article.tree.df<-cbind(article.ref,
                  article.pol,doc.length,assignments)
treemap(article.tree.df,index=c("assignments",'article.
ref'),
    vSize="doc.length",vColor="polarity", type="value",
    title="Guardan Articles mentioning Pakistan",
    palette=c("red","white","green"))
```

The treemap function relies on this data frame to construct Figure 5.21.

Guardan Articles mentioning Pakistan

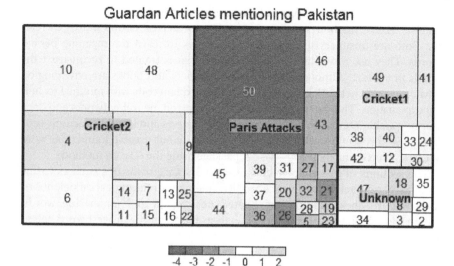

Figure 5.21 Illustrating the articles' size, polarity and topic grouping.

Step 6. Reach an insight or recommendation – In less than 100 lines of code you were able to execute a topic model and construct some exploratory visualizations. Following the six-step workflow, you are now able to reach an insight or conclusion. To begin this section, the problem was stated as "How does the Guardian newspaper prioritize articles about Pakistan?" Since the number of words in a particular article is represented as the square's area you can somewhat infer the amount of journalistic effort to create the article. Based on the limited analysis only one large article concerned the Paris attacks, ISIS and Syria, while the majority of writing in this timeframe concerned cricket. As expected, the language used to mention the Paris attacks was negative and cricket was mostly positive. Interestingly, the Paris attackers were part of the Islamic State of Iraq and Levant (ISIL), not Pakistani. However, the Guardian articles in this timeframe did mention Pakistan and were subsequently returned by the API.

5.4 Text to Vectors using text2vec

So far this chapter has sought to give you tools to identify underlying structure in text. These include document clustering, string distance calculations and topic modeling techniques. This last section examines a relatively new R package called `text2vec`. Text vectorization is either done using the `word2vec` or `GloVe` methods.

The word to vector (`word2vec`) was made public by Google engineers in 2013 and the `GloVe` method was pioneered at Stanford's NLP Group shortly thereafter. Both are unsupervised algorithms that seek to learn the meaning behind words. They use a two-layer neural network that is trained to reconstruct the words in context. Although many technical papers and books are emerging on covering neural networks, this section seeks to explain code with minimal technical explanation. The nuances of the lowly perceptron, neural networks and deep learning architectures are best left to in-depth courses and lengthy academic textbooks. So this section will give you the very basic but useable foundation word vectorizations using the `text2vec` R package using the `GloVe` method.

The sections of this chapter dealing with distance measures such as cosine similarity and Euclidean distance are all implementations of nearest neighbors. These approaches quantify the related documents and underlying words, which allows you to perform cluster analysis based on the nearest word neighbor. The distance measured is between a specific word and another specific word. However, these tools are limited for identifying a sophisticated word meaning, because the distance measure is singularly between terms. For example, the terms basketball and football may be close in many Euclidean distance measures because both are about sports. However, the terms represent two very different sports and have differing meaning, beyond being sports, such as the shape of the ball. So a more complicated method seeks to understand the differences between vectors of co-occurring words not individual distances between two words. Thus the co-occurrences of terms are calculated and represented as vectors. Then the multiple vectors are compared. This added complexity yields a more nuanced understanding of meanings between words.

For example, consider two documents made of only two words. The first is "text." The second is "mining." These can be represented in two dimensions as vectors shown in Figure 5.22. Now consider a new document with words "text mining." This vector of terms would lie in between the vector from the first two documents as shown.

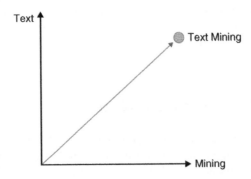

Figure 5.22 The vector space of single word documents and a third document sharing both terms.

While this is a crude example, it forms the foundation for a more complex vector space. Since documents contain many words they are represented in n-dimensions or hyperspace. Each word in a corpus gets its own vector dimension. Dimensions intersect when words appear together in the documents or co-occur.

5.4.1 Text2vec

The `text2vec` package provides fast implementations of a vector-based approach. It is used best with large corpora and with some term weighting like TFIDF or vocabulary pruning. In the previous chapter, you worked on 1000 randomly selected Airbnb reviews from the Boston area. Now we apply the text vectorization to a larger corpus of approximately 43000 Airbnb reviews.

The preprocessing of very large corpora needs to be done in an iterator which performs the task chunk by chunk on the text. The `text2vec` package provides an iterator function for this purpose. However, the example in this section is able to be executed on a single machine with only 4GB of RAM, so preprocessing is performed directly on the text vector.

Getting the Data

Please navigate to www.tedkwartler.com and follow the download link. For this analysis, please download "Airbnb-boston_only.csv." This file contains all Airbnb reviews for the Boston area.

Once you have the data stored locally, the script employs a new function for importing large files. The `fread` which the `data.table` package author says is a "fast and friendly file finagler." This is a convenient function for importing large files because it provides a percent of file read to track progress, handles mixed data types well and also reads the file quicker than `read.csv`.

```
library(data.table)
library(text2vec)
library(tm)
text<-fread('Airbnb-boston_only.csv')
```

The text object is both a data frame and a `data.table` object. The next line simply creates a new object called `airbnb`. This may seem redundant and computationally intensive but it is done to preserve the original text should you make a text cleaning error.

```
airbnb<-data.table(review_id=text$review_id,
        comments=text$comments,
        review_scores_rating=text$review_scores_rating)
```

Rather than wrap a custom function as you did in the past, all the cleaning steps are performed directly on the vector, one at a time. You could write a custom function, but this is a single vector called comments and the extra effort to create a custom function isn't warranted. By now the preprocessing functions should be familiar and commonplace.

```
airbnb$comments<-
removeWords(airbnb$comments,c(stopwords('en'),'Boston'))

airbnb$comments<- removePunctuation(airbnb$comments)
airbnb$comments<- stripWhitespace(airbnb$comments)
airbnb$comments<- removeNumbers(airbnb$comments)
airbnb$comments<- tolower(airbnb$comments)
```

The string split function is then applied to each of the 43000 comments. The resulting tokens object is a list with each element representing a unique comment. The elements are made of individual word tokens since the split was performed at spaces between words.

```
tokens <- strsplit(airbnb$comments, split = " ",
fixed = T)
```

The next step is to create a vocabulary using the `create_vocabulary` function from `text2vec`. To do so you iterate over the tokens object by creating n-grams. In this example, n-grams are shown as 1 to 1, meaning single words, but you could change it to capture a vocabulary of bi-grams (2,2) or mixing uni-gram and bi-grams (1,2). The n-gram tokenization is done chunk by chunk because of the iterating function `itoken` that is nested in the `create_vocabulary` function. Since `text2vec` is meant to work on large corpora, the package author kindly provides a progress bar that will increase as your computer computes the vocabulary.

```
vocab <- create_vocabulary(itoken(tokens),ngram = c(1,
1))
```

Next the vocabulary is pruned, allowing you to throw out frequent terms and infrequent terms. Here the code is pruning words that appear less than five times in the vocabulary.

```
vocab<- prune_vocabulary(vocab, term_count_min = 5)
```

In the end, the vocabulary is a list of terms, term counts and document counts. For example, the following code shows a small portion of the terms, term counts and documents. The result will be printed to your console but is shown in Table 5.6.

```
vocab[[1]][221:225]
```

Next you will create the term co-occurrence matrix (TCM). To begin, you create an iterator again. Specifically, your iterator will not act with a function like `tolower`, working chunk by chunk to change words, but on the actual individual document words themselves. This is passed to the next function `vocab_vectorizer` in the next line. The `vocab_vectorizer` creates a

Table 5.6 A portion of the Airbnb Reviews vocabulary.

Terms	Term count	Document count
arrived	2675	2535
prăpare	1	1
bedre	1	1
lujos	1	1
listening	18	18

function which is used in constructing the TCM. It defines parameters of the matrix build and is therefore a custom function like the `clean.corpus` function shown elsewhere in the book rather than a data object like a matrix, data frame or list. This function is passed the iterative word tokens, and some additional parameters. The `grow_dtm=F` is a logical statement telling the function not to build a document term matrix, but it could be changed to T if you wanted a DTM. Next the skip gram window defines the number of n-grams around the target word for calculating the vector word embedding. To learn more about skip grams, refer to the call-out "What is the skip gram method?" In this script, the value is 5 so that the neural network uses the prior and former five words around a target word. Finally the iterator and vectorizer functions are passed to the `create_tcm` function. This constructs a TCM using the words in the iterator so as not to overwhelm a computer's memory and with the specific parameters defined previously in the vectorizer object.

```
iter <- itoken(tokens)
vectorizer <- vocab_vectorizer(vocab, grow_dtm = F,
skip_grams_window = 5)
tcm <- create_tcm(iter, vectorizer)
```

The result is a square matrix because all words co-occur with at least some other word in the corpus. In this case, reviewing the structure of the TCM tells you that there are 10,838 words, because the dimensions are 10838 by 10838. Further, the dimension names include co-occurring words such as "dreamy" and "goose" for both rows and columns. The values within the matrix represent the probabilities of one word co-occurring with another.

```
str(tcm)
Formal class 'dgTMatrix' [package "Matrix"] with 6 slots
   ..@ i       : int [1:1557265] 6678 1284 7363 1284
6079 2568 5993 5993 7449 2654 …
```

```
..@ j        : int [1:1557265] 10328 3918 10752 3922
6852 7846 9933 9934 7691 4773 …
..@ Dim      : int [1:2] 10838 10838
..@ Dimnames:List of 2
.. ..$ : chr [1:10838] "dreamy" "cell" "goose"
"gigantic" …
.. ..$ : chr [1:10838] "dreamy" "cell" "goose"
"gigantic" …
..@ x        : num [1:1557265] 0.333 0.2 0.5 0.5
0.333 …
..@ factors : list()
```

What is the Skip Gram Method?

In building linguistic models mimicking grammar there are two methods. The continuous bag of words and the skip gram model. The code shown in this section uses the skip gram model.

Consider the following preprocessed sentence.

"after fouling out of the game steph curry complained to the referee and threw his mouthpiece in frustration"

Choosing a skip gram window of 1 would mean that each word's context is defined by the words on either side of the target word. For example, the words in the sentence "after" and "out" are the context for the word "fouling." This is done for each word and its corresponding context words. So the target and contextual data may look like Table 5.7.

Skip gram modeling tries to predict the context words based on the target word. For example if the target word was "curry" then the model will try to predict the likelihood of steph and then curry predicting complained. Since there is a professional basketball player named Stephan Curry, it is more likely to predict

Table 5.7 A portion of the example sentence target and context words with a window of 1.

Target	Preceding context word	Proceeding context word
fouling	after	out
out	fouling	of
of	out	the
the	of	game
game	the	steph
steph	game	curry
…	…	…

Table 5.8 A portion of the input and output relationships used in skip gram modeling.

Input	Output
fouling	after
fouling	out
out	fouling
out	of
of	out
of	the
the	of
the	game
game	the
game	steph
steph	game
steph	curry

---

that than red curry if the larger corpus contained more basketball than cooking themes. An example of the prediction schema for target and context words is shown in Table 5.8.

The input and output objective function is defined over the entire document or large batches of the document terms. As the model builds it computes the loss between the input and output against the input and some "noise" or contrasting words. For example, predicting the probability of steph from curry with a noisy example that is predicting steph from some other word such as cavaliers. While the Cavaliers may be elsewhere in a document concerning basketball, Stephan Curry never played for them and is likely not directly a 1 window skip gram input. So the loss between the two probabilities steph, curry and cavaliers, curry is computed. As this loss function is calculated for inputs, outputs and inputs to noise is done across the entire document for each word and the word vectors move in hyperspace. The neural network model eventually learns how to distinguish real words from the random noise words. Once this is accomplished the word vectors have a remarkable ability to understand the words in context.

The `term co-occurrence` function uses a window of 5 n-grams not one so it is more complicated and accurate than the simple example outlined here.

Now you have to fit a vector model based on the TCM. In this example, you are calculating a `GloVe` model. The model will first construct vectors with the TCM as the data followed by the parameters you specified. Next the desired dimensions of the word vectors are specified as 50, but they can be changed to increase the hyperspace dimensionality. Next x_max specifies the maximum number of co-occurrences as the weighting function. Increasing this parameter brings in more information thereby ensuring a more stable model. However, this increases the computation time needed. Next the learning rate is specified. This is a parameter that may improve results by explicitly stating a low value, but is not recommended because the glove model employs an optimization algorithm for this purpose automatically. GloVe uses a gradient descent algorithm to optimize learning rates automatically. Lastly the number of iterations specifies the number of times the optimization algorithm is applied. More specifically it is the number of times the AdaGrad algorithm is applied to the neural network weights, and each round is called an epoch. At each epoch, the AdaGrad algorithm adjusts the learning rate of the model without you having to do so manually. AdaGrad performs large updates for infrequent terms and conversely enacts small learning rate changes for frequent terms. This makes it well suited for sparse matrices, as is often the case for text. You will see the AdaGrad algorithm at work when you run the code as it prints a loss result for each epoch. You will see the epoch cost being minimized over each epoch but do not be tempted to increase the number so that the cost becomes zero. This will result in an inaccurate model.

```
fit.glove <- glove(tcm = tcm,
            word_vectors_size = 50,
            x_max = 10, learning_rate = 0.2,
            num_iters = 15)
```

Recall that the dimensions of the TCM were 10838 by 10838, corresponding to individual words. The fit.glove object's second element, fit.glove$word_vectors, contains the word vectors. In it there is another list which contains 10838 individual vectors with 50 numbers for each. These represent a vector with 50 values for each word in the TCM. It is these vectors that can be explored to understand meaning and context.

Before starting to explore the word vectors you have to add the two fit.glove vectors together and assign the words to each value using rownames instead of a numeric index. Next you square then square root the values for use later. The code below performs all of these tasks.

```
word.vectors <- fit.glove$word_vectors[[1]] +
   fit.glove$word_vectors[[2]]
rownames(word.vectors) <- rownames(tcm)
word.vec.norm <- sqrt(rowSums(word.vectors^2))
```

The first example of exploring the word vectors is done next. To begin you select the vector for "walk," subtract the vector for "disappointed" and then add the vector for "good." So the result should be representative of what a good walk without disappointment is within the Airbnb reviews. This was chosen because in chapter 4 the Airbnb word clouds showed that good reviews were within walking distance. So hopefully the resulting good.walks contextual vector will illustrate what specifically about the walks make them good.

```
good.walks <- word.vectors['walk', , drop = FALSE] -
   word.vectors['disappointed', , drop = FALSE] +
   word.vectors['good', , drop = FALSE]
```

The good.walks object is a matrix with one row and 50 values corresponding to its coordinates in the hyperspace. To make a meaning of the seemingly arbitrary numbers the text2vec package provides a cosine distance function. It is passed the matrix query – in this case good.walks. Then it must be given the matrix source containing all the word vectors. Lastly it is passed the normalized values that were created earlier. The resulting cosine distance matrix is a single row matrix with a length 10838. This represents the cosine distance between the good.walks and all other words in the vocabulary.

```
cos.dist <- text2vec:::cosine(good.walks,
                              word.vectors,
                              word.vec.norm)
```

Since you will only care about the top terms, the following code sorts and prints the top 10 terms according to cosine similarity. Table 5.9 shows the top terms. In it you can see that the algorithm has correctly understood the

Table 5.9 The top vector terms from good.walks.

Term	Cosine distance
Walk	0.8319209
Close	0.8245839
T	0.7779517
Subway	0.7727851
Downtown	0.7647843
Short	0.7556932
Walking	0.7542060
Also	0.7485843
Convenient	0.7419344
Good	0.7411359

meaning of good walks. They are short and convenient, close to downtown and subways and T commuter rail.

```
head(sort(cos.dist[1,], decreasing = T), 10)
```

You can easily switch the terms to explore within the code. Additionally in Chapter 4 the mention of dirty sinks was considered negative in the reviews. Now the code has been changed to explore what other words are part of the contextual meaning for dirty sinks.

```
dirty.sink <- word.vectors['sink', , drop = FALSE] -
    word.vectors['condition', , drop = FALSE] +
    word.vectors['dirty', , drop = FALSE]


cos.dist <- text2vec:::cosine(dirty.sink,
                              word.vectors,
                              word.vec.norm)
head(sort(cos.dist[1,], decreasing = T), 10)
```

The resulting ten words from this code are shown in Table 5.10. Part of the contextual meaning of a dirty sink is mentions of other items in an Airbnb review that can be dirty. So the algorithm has started to understand the meaning of dirty applied to nouns.

With other large corpora the meanings will vary. It is best to first do some rudimentary exploration, as was shown in chapter 4, so the vector exploration shown here is more informed. If you were an Airbnb host not only should you

Table 5.10 The top ten terms demonstrating the cosine distance of dirty and other nouns.

Term	Cosine distance
Sink	0.8305685
Dirty	0.7270638
Toilet	0.6514039
Dishes	0.6298118
Bathtub	0.5420831
Microwave	0.5365451
Kitchenette	0.5272143
Hairdryer	0.5244365
Plug	0.5239127
Dishwasher	0.5159356

clean your sink but also the other items in the Table 5.10! There are more applications of text vectorization in terms of both identifying uncommon contextual meaning and predictive modeling.

5.5 Summary

In this chapter you applied various techniques to understand the underlying structure of resumes and newspaper articles. Specifically you learned:

- how to perform k-means clustering
- how to perform k-mediod clustering
- to use the `StringDist` library with `Hclust`
- what LDA is
- LDA topic modeling using LDA and LDAvis
- other topic modeling packages
- use `word2Vec` to get text vector calculations
- make a compelling treemap visualization

6

Document Classification: Finding Clickbait from Headlines

In this chapter, you'll learn

- how to create a DTM from a training set and use it when constructing new DTMs for new documents
- what a sparse model matrix is
- how to create a data partition for training and testing of the document classification algorithm
- how to create a document classification algorithm
- what lasso regression is and how it differs from linear regression
- perform cross validated lasso regression
- how to calculate the area under the curve (AUC) for the document classifier
- how to calculate the overall accuracy of the classifier on unseen headlines
- how to identify the words most contributing to clickbait headlines

6.1   What is Document Classification?

Document classification falls within the field of machine learning. Machine learning is an extension of artificial intelligence; it represents the tool set, methods and approaches allowing a computer to write code from data on behalf of the programmer.

There are two approaches to machine learning. The first set of algorithms make up "unsupervised learning." Without a dependent or outcome variable the unsupervised learning algorithms look for complex patterns within the data and infer outcomes such as "cluster." Unsupervised approaches were explored in the last chapter. For review, if you applied unsupervised approaches to all historical basketball teams with seasonal stats, an algorithm could identify the best and worst team clusters without a pre-existing label "best" or "worst."

Text Mining in Practice with R, First Edition. Ted Kwartler.
© 2017 John Wiley & Sons Ltd. Published 2017 by John Wiley & Sons Ltd.

However, in this chapter you explore approaches for supervised learning. Supervised learning algorithms require an outcome or "y" variable for each observation. Supervised learning methods require the defined outcome for each observation or document. The result of the algorithm is not a cluster, but a prediction or classification for "y." The prediction is the algorithm's best guess at the "y" value. Figure 6.1 illustrates the development step of taking labeled "y" data and training an algorithm. Figure 6.2 shows how new data is given to the algorithm and the outcome is then predicted.

Drilling down further, there are two types of outcome variables for supervised learning. The first is classification and the second is continuous prediction. For example, predicting the winner of a basketball game can be done using classification or prediction. A team can win or lose a game. These represent two classes, win or lose, for the outcome and as such you could build a classification algorithm. In contrast, you could build a different algorithm to predict the team scores. Predicting scores is a continuous outcome because the points have a continuous range between 0 and some larger number.

This chapter covers classification outcomes related to documents. The model inputs are document terms, or term weights. The algorithm is then applied to new documents so that the new documents' category is classified. The next chapter uses text to make classifications or predictions too. However, the output is not related to another new document. Instead the output in the next chapter's algorithms is an external prediction or classification. For the purposes of this chapter, a simple document classification definition is below.

Document classification assigns a document to one or more classes or categories.

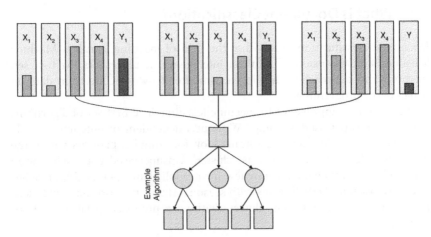

Figure 6.1 The training step where labeled data is fed into the algorithm.

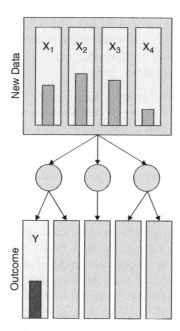

Figure 6.2 The algorithm is applied to new data and the output is a predicted value or class.

There are many machine learning books devoted to in-depth explanation and analysis. This book is meant to be a practical guide for text mining. As such, this chapter demonstrates a case study and one method for document classification. The principles learned in this chapter can aid in applying different algorithms and in other use cases. For example, a well-documented approach is to use naive Bayes for spam classification although that is not the method illustrated in this chapter. You can extend your knowledge beyond this case study by reading specific machine learning books.

6.2 Clickbait Case Study

In August 2016, leaders at Facebook announced a plan to identify and limit clickbait, because the Facebook newsfeed goal is to "show people the stories most relevant to them." Facebook leaders contend that clickbait clutters newsfeeds and detracts from authentic communication. The Facebook algorithm scores headlines according to whether or not the headline *withholds* information (requiring the user to click the link) and whether or not the text is *misleading* or exaggerated. Clickbait links often have both of these attributes. According the Facebook announcement limiting clickbait, the document classification algorithm was trained on tens of thousands of headlines according to these two dimensions. For simplicity, you will take a simpler approach classifying a smaller number of headlines as clickbait.

Examples of clickbait include

- The Most Shocking 'Jerry Springer' Episode EVER!!
- A Man Spots A Reckless Driver, When He Stops Her I'm Shocked. OMG!!
- ALERT: Diet Pepsi Removes Aspartame And Replaces It With An Equally Dangerous Chemical!
- How To Get Weed In 5 Minutes Without Leaving Your Couch
- She Was Born With Birthmarks That Covered Her Entire Body But Wait Until You See Her Now!

In this case study, suppose you are a text miner working at a large social media company. The company makes money through ads placed in and around its newsfeed. To remain salient, the feed needs to have engaging content. However, dubious social posts are cluttering the newsfeed, thereby affecting the ad revenue. These clickbait posts lure your users away from the newsfeed, costing the company ad revenue. Company leaders have asked you to create a simple algorithm to identify clickbait posts so they can be removed from the newsfeed. Your text mining workflow may look something similar to this.

1) **Define the problem and specific goals.** Create a document classification algorithm identifying clickbait. Evaluate the algorithm and make suggestions on improving its performance.
2) **Identify the text that needs to be collected.** For this example, your model will be built using 3000 headlines. The "y" variable is 1 if the headline came from a clickbait source and 0 if the source was considered newsworthy.

Getting the Data

Please navigate to *www.tedkwartler.com* and follow the download link. For this analysis, please download "all_3k_headlines.csv." The file headlines were obtained from multiple news and clickbait sources. The file contains three columns, headline, url and source.

---

3) **Organize the text.** To begin, the text will be separated into training and test sets. The model will be built using the training, and evaluated based on the test set.
4) **Extract features.** To build a classification algorithm you will construct a TFIDF weighted DTM from the training data. The original DTM matrix TFIDF terms and weight will be used to construct the test DTM. This allows you to train the model separately without preprocessing the entire corpus when presented with new unseen documents.

5) **Analyze. DTM matrices are extremely sparse.** As a result, you will train a GLMNet lasso regression. Lasso regression is memory efficient, making it ideal for document classification. Then you will evaluate the classification algorithm's accuracy and determine the most impactful clickbait terms.

6) **Reach an insight or recommendation.** You will know the top terms signaling clickbait. Further you will be able to recommend ways to improve the algorithm's accuracy.

6.2.1 Session and Data Set-Up

After loading the headlines data frame into your R session you need to add the appropriate packages. The tm library is very familiar by now. Matrix provides memory efficient methods for manipulating sparse matrices. Recall, DTMs are sparse, containing many thousands of columns made primarily of zeros. The glmnet package provides elastic net regression which is a combination of lasso and ridge regression. The regression specifics will be explained later, but this package provides the Fortran bindings for training and predictions. The caret library is useful for data partitioning, categorization and regression training. The pROC library helps create and analyze receiver operator curves (ROC). The ROC is an evaluation metric used in classification problems. The last two packages, ggplot2 and ggthemes, are used for visualizations.

```
library(tm)
library(Matrix)
library(glmnet)
library(caret)
library(pROC)
library(ggthemes)
library(ggplot2)
library(arm)
```

In order to have consistent preprocessing, it is best to create a custom function. Then you can apply the same steps to the training data and on any new documents that need to be classified. Usually headlines do not have misspellings, emoji or unusual characters so the function is very basic.

```
headline.clean<-function(x){
  x<-tolower(x)
  x<-removeWords(x,stopwords('en'))
  x<-removePunctuation(x)
  x<-stripWhitespace(x)
  return(x)
}
```

To avoid recreating the entire DTM and retraining a model each time you are presented with new documents you must match a new document's attributes to the existing DTM terms that were used for model training. For example, if you create a model based on 1000 terms represented as matrix columns, any new document must have a value for the same 1000 terms. That means the new document DTM may have to lose terms that were not part of the original data or original DTM terms to the new DTM. To score a new document, the algorithm will be expecting the new DTM to have the same number of columns for new data as was contained in the training data.

The RTextTools library contains a function called create_matrix which allows you to reference an originalMatrix for constructing a new DTM. However, the library is not actively maintained and contains a typo in line 42 of the code. If you decide to use the create_matrix function you will need to fix the original code using this snippet. Once the trace window appears navigate to line 42 and change "A" to "a." Figure 6.3 is a screenshot of the trace window showing the capital "A" that needs to be changed at line 42.

```
install.packages("RTextTools")
library(RTextTools)
trace("create_matrix",edit=T)
```

To avoid using an unmaintained package, you can use the simpler version of create_matrix shown below, called match.matrix. The custom match.matrix function drops the internal preprocessing steps and tokenization from create_matrix. The match.matrix function retains the ability to reference an original matrix for new document DTMs. This will be shown later in the code.

match.matrix accepts a single text column, an original matrix object if desired and a term weighting parameter. Recall that DTMs can be constructed using term frequency or TFIDF weighting. You can specify either when calling this function. Within this function the DTM controls are defined first. Next, the vector's encoding is changed to avoid errors stemming from unusual characters. After that, the VCorpus function is applied to the text vector. A DTM is constructed with DocumentTermMatrix along with the previous control input.

```
38   if (!is.null(originalMatrix)) {
39     terms <- colnames(originalMatrix[, which(!colnames(originalMatrix) %in%
40       colnames(matrix))])
41     weight <- 0
42     if (attr(weighting, "Acronym") == "tf-idf")
43       weight <- 1e-09
```

Figure 6.3 Line 42 of the trace window where "Acronym" needs to be changed to "acronym."

The next internal function code beginning with an `if` statement retains the original matrix's information to be applied to a new document. If the `original.matrix` parameter is not `NULL` then the matching will occur. This is a programmatic way of saying "if you reference an original matrix then begin the matching process and if not then ignore the matching process." Specifically the terms in the new matrix that are not in the original are dropped. Any terms in the original that are not in the new documents are added. The weight of all new terms is set at 0 so that it does not impact the original DTM weights. There is a second `if` statement in case the weighting is `tf-idf`. If the original DTM weighting is `tf-idf` then the weight is changed to a near zero amount, `0.000000001`, to avoid an error. Ultimately, the output of this function is a DTM, as you have used many times throughout this book. When constructing the training matrix you would not add an `original.matrix` parameter but later during testing on new documents you would reference the original matrix object for matching and weighting.

```
match.matrix <- function(text.col,
                   original.matrix=NULL,
                   weighting=weightTf)
{
  control <- list(weighting=weighting)
  training.col <-
    sapply(as.vector(text.col,mode="character"),iconv,
       to="UTF8",sub="byte")
  corpus <- VCorpus(VectorSource(training.col))
  matrix <- DocumentTermMatrix(corpus,control=control);
  if (!is.null(original.matrix)) {
    terms <-
     colnames(original.matrix[,
which(!colnames(original.matrix) %in% colnames(matrix))])
    weight <- 0
    if (attr(original.matrix,"weighting")[2] =="tfidf")
      weight <- 0.000000001
    amat <- matrix(weight,nrow=nrow(matrix),
             ncol=length(terms))
    colnames(amat) <- terms
    rownames(amat) <- rownames(matrix)
    fixed <- as.DocumentTermMatrix(
      cbind(matrix[,which(colnames(matrix) %in%
                   colnames(original.matrix))],amat),
       weighting=weighting)
    matrix <- fixed
  }
```

```
matrix <- matrix[,sort(colnames(matrix))]
gc()
return(matrix)
}
  }
```

With the set-up functions and libraries loaded you are now able to start training your document classification algorithm. With the set-up and custom functions complete you have completed step 3, text organization, of the text mining workflow.

6.2.2   GLMNet Training

The next step in the text mining workflow is to extract features. To do so, parse the `all_3k_headlines.csv` file. The `createDataPartition` function selects random number rows based on the dependent variable distribution. This is helpful if you have unbalanced classes or multiple classes. The `p` parameter will select 50% of all row numbers. This leaves 50% or 1500 headlines to be used as a holdout test set. You can adjust this parameter to increase the number of records used in the training split.

When building models it is important to have a holdout set. Otherwise you run the risk of "overfitting" your model. Without a holdout, you may believe the model is vastly more accurate than it really is when confronted with new data. In this snippet, the `train` number vector is used to index the headlines data into `train.headlines` and using a minus sign, removing the `train` numbers leaving the `test.headlines`.

```
headlines<-read.csv('all_3k_headlines.csv')
train<-createDataPartition(headlines$y,p=0.5,list=F)
train.headlines<-headlines[train,]
test.headlines<-headlines[-train,]
```

At this point, the code will focus on the training data. First apply the `headline.clean` function to preprocess the text vector. Then apply the `match.matrix` function to construct a DTM. Within `match.matrix` add the term weighting `tm::weightTfIdf`. The double colon tells R to use the specific meaning of `weightTfIdf` from the tm "namespace." If you want to weight by term frequency, use `tm::weightTf`. You do not need to specify the third parameter, `original.matrix`, since `train.dtm` *is* the original matrix

```
clean.train<-headline.clean(train.headlines$headline)
train.dtm <- match.matrix(clean.train,
weighting=tm::weightTfIdf)
```

The resulting `train.dtm` object is a familiar DTM from the `tm` package. Calling `train.dtm` provides information including the 4996 terms among the 1500 documents.

```
> train.dtm
<<DocumentTermMatrix (documents: 1500, terms: 4996)>>
Non-/sparse entries: 10461/7483539
Sparsity           : 100%
Maximal term length: 23
Weighting          : term frequency - inverse document
frequency (normalized) (tf-idf)
```

The DTM needs to be changed into an acceptable object for modeling. First change the DTM to a simple matrix with `as.matrix`. Then using `Matrix` change the object into a sparse matrix. The sparse matrix drops the zeros completely, making the object very memory efficient. Within `Matrix`, the `sparse=T` parameter automatically becomes TRUE if more than half the values are 0. The parameter is shown here for educational purposes only.

```
train.matrix<-as.matrix(train.dtm)
train.matrix<-Matrix(train.matrix, sparse=T)
```

To understand the differences between `train.dtm` and the sparse `train.matrix` use the code below. The dimensions are still 1500 rows by 4996 columns. However, when calling a section of the `train.matrix` the zero values are now a period. This makes the matrix very lightweight for modeling.

```
dim(train.matrix)
train.matrix[1:5,1:25]
```

What are ElasticNet, Ridge and Lasso Regression?

Foundationally ElasticNet, ridge and lasso approaches are forms of multiple variable regression. According to Barry Keating in *Business Forecasting with ForecastX* "multiple regression is a statistical procedure in which a dependent variable is modeled as a function of more than one independent variable." That is to say the outcome variable is predicted by a linear combination of unrelated inputs.

For example, to predict the number of ice cream cones sold, a linear regression may have inputs including "number of people passing by" and "temperature." The linear regression model may use a y intercept, called "beta-naught (β_0)," and beta coefficients for the "number of people passing by" and "price" inputs. These input coefficients are represented as "beta-1 (β_1)" and "beta-2 (β_2)"

respectively. With more independent variables or inputs comes more beta coefficients. Lastly, the regression has an error term to account for errors between predicted and actual values. With this information, the ice cream cone regression could be written as:

$$Number\ of\ Ice\ Cream\ Cones = \beta_0 + \beta_1 X_1 + \beta_2 X_2 + \varepsilon$$

Where X_1 is equal to the number of people passing by and X_2 equals the temperature. If the linear model was created, the beta coefficients would contain numeric information affecting the baseline ice cream cones sold, β_0. In this example, the equation may look like:

$$Number\ of\ Ice\ Cream\ Cones = 15 + (.1 * \# of\ people) + (-.20 * price)$$

Reviewing the coefficients, the baseline number of ice cream cones sold is 15. For each 10 people passing by, you can expect to sell an additional ice cream cone (0.1*10). That number is offset by the negative coefficient assigned to price. In this fictitious example, if the price was set at $5 per cone, sales dip by 1 cone (−0.2*5.). If the ice cream stand had 100 people pass by while prices were $5 per cone, the stand would sell 24 cones. The linear combination of inputs is 15 + 10 (0.1*100) − 1 (−0.2*5). As you build a regression model keep in mind that coefficients should align with expectations about the business problem. For example, using temperature as an additional factor should increase the ice cream sales and the resulting coefficient should be positive.

In the ice cream example, the outcome, number of ice cream cones, is continuous. However, a special type of regression called logistic regression can be used to calculate the logit, which can be changed into probability between 0 and 1. Since the probability outcome is between 0 and 1 a logistic regression can be used for binary classification.

The `glmnet` package constructs logistic regression among other types. The "glm" stands for "generalized linear model." Generalized linear models are more flexible than ordinary linear regression models because GLMs allow for response variable errors to be non-normally distributed. Additionally, the `glmnet` models employ a penalty parameter. Specifically, the penalty parameters are employed in both ridge and lasso regression. The penalty variable is called "lambda." The mixture between lasso and ridge is captured in the "alpha" parameter. When `alpha=1` the package performs lasso. Conversely when `alpha = 0`, ridge regression is performed. Alpha values between 0 and 1 mix the penalty parameter method resulting in an ElasticNet model.

Lasso stands for "least absolute shrinkage selection operator." Lasso performs variable selection with regularization by ignoring inputs that have little explanatory power. The lasso method forces the sum of the beta coefficients to be less than a fixed amount. Therefore certain coefficients are forced to 0. The result is a smaller model with fewer inputs, which is easily interpreted. This method is great for text classification because the training data contains so many columns. Rather than create many small coefficients for thousands of columns (terms), the input words are penalized. This forces the model to throw out many thousands of words and, in doing so, the model is constructed accurately and quickly. For lasso regression the regularization or penalty calculation method is "L1."

The ridge regression technique adds a quadratic part to lasso's regularization penalty. Similar to lasso, ridge regression assumes a constant variance and input variable independence. However, the regularization calculation in ridge regression is "L2."

In both L1 and L2 regularization the calculations penalize document terms or model inputs. By removing many inputs to a model the hope is that the simpler algorithm is more applicable to new data and avoids overfitting. In both cases the penalty ensures that large coefficients are not assigned to rare words. Rare words with large coefficients reduce the effectiveness of the model because new documents are unlikely to contain the rare word. Using regularization avoids assigning coefficients to rare terms thereby improving the model's accuracy for new observations. The math underlying L1 and L2 calculations requires significant explanation beyond the scope of this small section. In *very* short, L1 and L2 are two methods for calculating the magnitude of a value using a jagged geometric distance versus a straight line. An analogy would be measuring distance "as the crow flies" or navigating city blocks in a jagged pattern between two points.

For practical purposes, remember that lasso regression using L1 regularization attempts to change the coefficients to zero in as many input variables as possible. In contrast, ridge regression's L2 regularization will push coefficients towards zero but not exactly zero. Generally speaking, lasso will be a simpler model because it ignores more inputs than ridge but often lasso is slightly less accurate than ridge. This is because the L2 regularization is able to keep more inputs compared to L1's need to force inputs to zero.

Finally, mixing the two regularization techniques by changing alpha between 0 and 1 will create an elastic net regression model. The elastic net regression is a linear combination of L1 and L2 penalty methods. In the end, the alpha input is a model tuning parameter balancing accuracy, and simplicity.

You are now ready to train your `glmnet` model using `cv.glmnet`. The `cv.glmnet` function performs cross validation when constructing a `glmnet` model. Cross-validation is a method for breaking up the training data and repeatedly shuffling the groups. Iteratively training on different groups of data

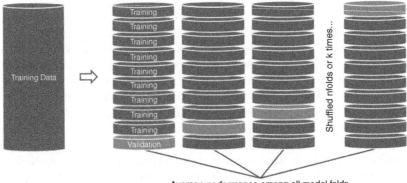

Average performance among all model folds.

Figure 6.4 Training data is split into nine training sections, with one holdout portion used to evaluate the classifier's performance. The process is repeated while shuffling the holdout partitions. The resulting performance measures from each of the models are averaged so the classifier is more reliable.

helps reduce the likelihood of overfitting. As a result, the model has a better chance of generalizing to new data when put into practice. At the conclusion of the multiple model builds, the classifier performance measures are averaged. In cross validation, every record has a turn at being in the training and then in the holdout set. Figure 6.4 visually depicts cross validation on a training data set.

To create the object `cv`, pass the sparse matrix `train.matrix` into the `cv.glmnet` function. Next define the outcome or `y` variable. The sparse matrix contains the word columns but not the 0 or 1 outcome variable. Thus the `y=` parameter uses the column `train.headlines$y` from the original training set. Further, the column contains numbers and must be changed to a categorical variable using `as.factor`. Next, the `alpha` parameter is the mixing input for how regularization (penalty) is calculated. If interested see the prior section for more detail. As written `alpha=1` will ensure that a lasso regression model is created. The `family=` parameter defines how the model should behave. The complete family parameter options include "Gaussian," "binomial," "Poisson," "multinomial," "cox" and "mgaussian." Since the outcome "clickbait" is binary, the family is `binomial` forcing the model to build a logistic regression. If the outcome has multiple classes, this parameter would change to `multinomial`. The `nfolds` input defines the number of cross validation folds to perform. The higher the number, the longer computational time, but the model has an improved chance of being reliable when put to use. The last parameter `type.measure='class'` selects the best penalty "lambda" value from among the cross validation models based on the lowest misclassification rate between clickbait and legitimate headlines. Other measures can be chosen depending on the model family and the misclassification implications. The `type.measure` inputs include "deviance," "AUC," "class," "mse" and "mae."

```
cv<-cv.glmnet(train.matrix,
y=as.factor(train.headlines$y), alpha=1,
family='binomial', nfolds=10, intercept=F,
type.measure = 'class')
```

The `cv` object is a list containing the model information. One way to review the outcome is by calling `plot` on the model object. The graphic demonstrates the relationship between the misclassification rate and the lambda penalties. Changing the `type.measure` parameter when calling `cv.glmnet` will change the y axis accordingly. Within the graph, there are two vertical dotted lines. The left-most line represents the lambda value which minimizes the misclassification rate. The other dotted vertical line represents the highest regularization value within one standard deviation of the minimal class error. A model based on second penalty value will have fewer inputs compared to a model using the first dotted line lambda value. This helps you make an informed decision to balance complexity against accuracy when making predictions. Figure 6.5 demonstrates the cross validation results visually.

```
plot(cv)
```

Next use the classifier model on the training headlines to gain an understanding of the model's accuracy. Keep in mind that the classifier is now being applied to headlines that it has already seen, and the accuracy will be inflated. Still, it is a good idea to review this information before applying the classifier to the test data. If the model does not perform well here, investigate any data integrity issues, change the training set size, adjust text preprocessing and finally adjust the tuning parameters of the lasso regression. Doing so before applying the model to the test set ensures an a priori approach to the classification construction.

Use the `predict` function to apply the lasso regression to data. First, pass in the lasso regression, `cv`, and a matrix of inputs. The matrix must contain the

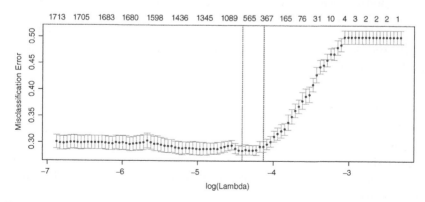

Figure 6.5 The interaction between lambda values and misclassification rates for the lasso regression model.

same columns as was used during the training step. Next the `type` parameter specifies whether or not the returned information should be the class "1" or "0." Changing this parameter to `type="response"` will return the probabilities instead of the final class for each row. Lastly, `s=cv$lambda.1se` instructs the `predict` function to use a lambda value corresponding to the second vertical dotted line in the previous figure. This minimizes the inputs without losing a lot of accuracy. You can change the `s` value to `cv$lambda.min` to use a more complex model that truly minimizes the misclassification rate.

```
preds<-predict(cv,train.matrix,type='class',
          s=cv$lambda.1se)
```

The `preds` object is a matrix with 1500 rows and a single column. The values correspond to 0 for a legitimate headline and 1 for clickbait. If probabilities are needed, change the type to `response` in the previous code.

Using the `roc` function, you can calculate the receiver operator characteristics (ROC) curve. The curve represents the relationship between how sensitive and specific the model is. In other words, the relationship between the true positive rate on the y axis and the false positive rate on the x axis is displayed. For example, if you were a doctor trying to classify cancerous tumors, you may want to be more sensitive and less specific. In doing so, you would lower the threshold of classifying a malignant tumor at the expense of being accurate. This would mean more false positives, but would likely give you a better chance of saving lives, because you would address more tumors in total. In contrast, a model classifying citizens as terrorists must be very specific. Law enforcement cannot arrest large swaths of the population without some backlash. As a result, a terrorist classification has a high threshold and is not very sensitive. Instead, it would be very specific and require a high probability of being correct.

The total area under the curve (AUC) is a performance indicator for classification models. An AUC of 0.5 means that the model is no better than a random guess. AUC values below 0.5 are worse than random. A perfect model would have an AUC equal to 1. Use the code snippet to calculate the ROC and AUC for the training predictions. For `roc`, pass in the original classes and then the predicted values.

```
train.auc<-roc(train.headlines$y,as.numeric(preds))
train.auc
```

Calling `train.auc` in your console will print data characteristics and the area under the curve result. An AUC of 0.892 is more than 0.5 and is approaching 1 so the model is performing well. Keep in mind that the score represents predictions on headlines that the model has already seen. Thus, it will perform better then predictions applied to the test headline set.

```
> train.auc
```

```
Call:
roc.default(response = train.headlines$y, predictor =
as.numeric(preds))
```

```
Data: as.numeric(preds) in 750 controls (train.
headlines$y 0) < 750 cases (train.headlines$y 1).
Area under the curve: 0.892
```

Call `plot` on the `train.auc` object to create the ROC curve. The diagonal line at 45° represents a random guess. The area under the diagonal line is 50% of the total square so it represents a 50/50 guess. The line extending upwards represents the ROC for the training set predictions. The line extends upwards above the diagonal line. The visualization is shown in Figure 6.6.

```
plot(train.auc)
```

Another way to review the performance is to review the classification accuracy. This can be done using a confusion matrix. To tally the results of the predictions and the actual headline values use the `table` function. In the resulting `confusion` matrix, rows represent the predictions from the model while the columns are the actual headline classes in the training set. For this model the confusion matrix results are presented in Table 6.1.

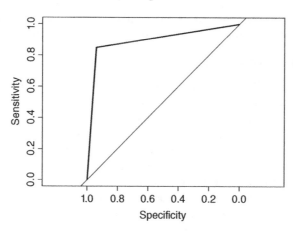

Figure 6.6 The ROC using the lasso regression predictions applied to the training headlines.

Table 6.1 The confusion table for the training set.

		Actual classes	
		0 = legitimate headlines	1 = clickbait
Predicted classes	0 = legitimate headlines	705	117
	1 = clickbait	45	633

```
confusion<-table(preds,train.headlines$y)
```

The intersections of the same classes denote the headlines that the model predicted correctly. Conversely the class differences represent incorrect predictions. So the model predicted a legitimate headline 822 (705 + 117) times. The model was correct 705 times and incorrect on 117 other headlines. The same is true for the clickbait class. The total clickbait predictions were 678. This time the intersection shows 633 correct classifications versus 45 incorrect. To calculate overall model accuracy, sum the diagonal, representing the correct predictions for both classes, then divide by all observations. The code below uses the `diag` function nested inside the `sum` function and then divides that amount by the matrix sum. `Diag` will extract all values along the diagonal, no matter what the size of the matrix. This vector is then summed. In this case, the model classified 89.2% (1338/1500) of headlines correctly. Your model may differ slightly because the randomness of the data partition.

```
sum(diag(confusion))/sum(confusion)
```

So far, the analysis of the model is based on training data. You are technically analyzing the model's output to ensure its overall accuracy and consistency. Within the text mining workflow model building and review completes the information extraction, step 4. The extracted information, a classification model that has been reviewed, is now applied to the test data. Applying the model to test data represents step 5 in the text mining workflow.

6.2.3 GLMNet Test Predictions

With your cross validated model constructed you are now able to make predictions on new headlines. To do so, the model needs a matrix containing values with the same weighting as the training data, along with matching column names. Applying the `headline.clean` function to the test set ensures preprocessing consistency. Next, apply the `match.matrix` function to the test headlines. As before pass in the headline object then the weighting parameter. Be sure to use the same weighting as before. Lastly, pass in a parameter for `original.matrix`. Passing in the training matrix, `train.dtm`, ensures that the columns match between the training and test matrices. Specifically, test words that do not appear in the training set are dropped. The words in common are retained, and columns of zeros are appended for any words that were in the training set but not in the test set. This is done because it is impossible for the lasso regression to make a classification on information it has not already seen, such as new words in the test set. Further, the regression model is expecting to have a value, even zero, for all columns presented during training.

```
clean.test<-headline.clean(test.headlines$headline)
test.dtm<-match.matrix(clean.test,
                weighting=tm::weightTfIdf,
                original.matrix=train.dtm)
```

test.dtm is another DTM from the tm package. Compare the previous train.dtm and test.dtm by calling the objects in the console. Both DTMs contain 4996 terms. The data partition split the headlines in half, so both contain 1500 documents, but this is not necessary. The only requirement is that the regression inputs must match, *not* the number of observations. The code below calls both test.dtm and train.dtm. The only difference between the two is the number of non-sparse terms.

```
> test.dtm
<<DocumentTermMatrix (documents: 1500, terms: 4996)>>
Non-/sparse entries: 7273/7486727
Sparsity            : 100%
Maximal term length: 23
Weighting           : term frequency - inverse document
frequency (normalized) (tf-idf)

> train.dtm
<<DocumentTermMatrix (documents: 1500, terms: 4996)>>
Non-/sparse entries: 10461/7483539
Sparsity            : 100%
Maximal term length: 23
Weighting           : term frequency - inverse document
frequency (normalized) (tf-idf)
```

As before, change the test.dtm to a simple matrix. Then change the object to the memory efficient sparse matrix with Matrix. You can nest these functions into a single line or use the pipe operator to make the code more concise. The code below separates the functions for educational purposes, rather than nest functions or use the pipe operator.

```
test.matrix<-as.matrix(test.dtm)
test.matrix<-Matrix(test.matrix)
```

To make predictions on the test data, use the predict function with the model object. The predict function accepts the cv model first. Then the response type "class" is specified. With "class" the class predictions, 1 or 0, are returned. In contrast, using type ="response" will produce the probability for each of the test headlines. Lastly, the "s" parameter is defined. In Figure 6.6, there were two lambda values denoted by dotted lines. The first represented the lambda value that minimized the overall misclassification error. The right-hand lambda value in Figure 6.6 maximized the penalty within one standard deviation of the minimum or best performing model. This value has the lowest number of inputs, and sacrifices a little accuracy. The code below specifies the minimum lambda value from the cross validated model. Within the cv object, the value is captured as cv$lambda.min. To drop more inputs, thereby making the model simpler yet retain most of the predictive accuracy, specify `s=cv$lambda.1se`.

```
preds<-predict(cv,test.matrix,type='class',
              s=cv$lambda.min)
headline.preds<-data.frame(doc_row =
       rownames(test.headlines),class=preds[,1])
```

The second line creates a concise data frame with appropriate column names and referencing the test rows from the original data set. To examine the class predictions you can call `head` on the preds. Within the table, class is a categorical factor and must be changed to numeric for calculating the ROC. This information is given in Table 6.2.

6.2.4 Test Set Evaluation

The `preds` object is a matrix with 1500 results. Evaluation metrics can be calculated on the test set because the test data represents known headlines and outcomes. When the model is applied to truly new headlines, outcomes are unknown so calculating the ROC is not possible. In practical application, periodic sampling and classification reviews help to keep the model from becoming stale. Using labeled test data, the `roc` function calculates the receiver operator characteristics curve as before. Pass in the actual test headline classes with the predicted classes after the predictions are changed to numeric. Calling the `test.auc` object prints the AUC information.

```
test.auc<-roc(test.headlines$y,as.numeric(preds))
test.auc

> test.auc

Call:
```

Table 6.2 The first six rows from `headline.preds`.

doc_row	class
1	1
2	1
3	1
6	1
7	0
9	0

```
roc.default(response = test.headlines$y, predictor =
as.numeric(preds))
```

```
Data: as.numeric(preds) in 750 controls (test.
headlines$y 0) < 750 cases (test.headlines$y 1).
Area under the curve: 0.7698
```

You should expect a lower AUC because these headlines represent completely new and unseen information compared to the cross validated training set. In this case, the `train.auc` value is 0.892 compared to the `test.auc` value of 0.7698. This drop-off is not significant, and both are above 0.75. Therefore the approach is showing merit at identifying clickbait headlines. AUC values above 0.5 mean that the model is better than random guessing. A value of 1.0 means that the model is likely overfitting the data and has some flaw. To improve this model's `test.auc` score a data scientist or text miner could increase the number of training headlines, add bi-gram and tri-grams to the matrices making both wider, or increase the number of headline sources to add diversity to the training set.

Use the code below to visually compare the AUC curves for both models. Using the base `plot` device function, pass in the `train.auc` specifying the color "blue" along with a title using the `main` parameter. Next add a layer for the `test.auc` curve to be plotted in "red." The code below employs `lty=2` so that the line is dashed to add further contrast. As you move along the test set AUC curve it is lower than the training set yet remains above the diagonal random threshold line. Keep in mind that depending on the data partition the exact AUC and resulting visual may differ slightly. Figure 6.7 compares both AUC curves.

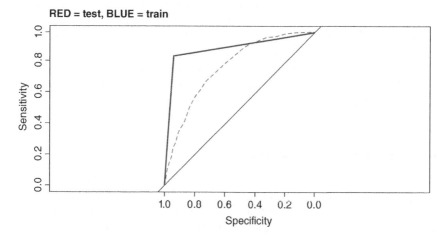

Figure 6.7 A comparison between training and test sets.

```
plot(train.auc,col="blue",main="RED = test, BLUE =
train",adj=0)
plot(test.auc, add=TRUE,col="red", lty=2)
```

Next use `table` to create a confusion matrix. The function accepts the predictions contained in the second column of the `headline.preds` data frame along with the actual outcomes from the test set, `test.headlines$y`. In the second line the sum of the diagonal divided by the total cases calculates the misclassification rate for the test set. It is likely the case that rerunning the model will vary because of the data partition and the way the lasso regression removes inputs based penalty calculations.

```
confusion<-table(headline.preds[,2],test.headlines$y)
sum(diag(confusion))/sum(confusion)
```

A Word of Caution

Avoid repeatedly rerunning the test score analysis hoping for improved results based solely on partitioning and lasso penalty behavior. When the test set evaluation is rerun, accuracy scores have a range of 0.70 to 0.76. Rerunning the analysis in the hopes of a better partition violates a priori assumptions. Fishing for scores based on a random split is not sound data science. Instead, make final adjustments to tuning parameters when training the model not after seeing the test set results. At the training stage, adjustments should be made to cross validation folds, the size of the training set itself and the alpha parameter. In fact, it is a good idea to rerun the model building step with training data using `alpha=0`. This will change the regression from lasso to ridge. Then retrain the model using an alpha between 0 and 1. This will help you learn the accuracy tradeoffs between lasso, ridge and elastic net regression. For each, the accuracy will change along with this tuning parameter. Once you feel comfortable that the results are maximized to your best a priori knowledge and the model is stable with a large number of cross validation folds then proceed to apply the tuned model to the test set. The test set scores are the best indication of how the model will behave when put to real use.

6.2.5 Finding the Most Impactful Words

Analyzing both evaluation metrics and individual model inputs is needed to complete step 5 of the text mining workflow. A benefit of using linear regression is that the model can be examined for insights. The beta coefficients have a magnitude and positive or negative sign relating to the outcome relationship. A text classification model can have thousands of inputs which can be overwhelming to process. Instead, only reviewing the top and bottom terms can still be meaningful.

To organize the coefficients for the thousands of term inputs, the `coef` function is applied to the `cv` linear model object along with a specific lambda value. Doing so extracts the coefficient values for each input term, and the input terms are captured as row names. The numeric information is returned as a sparse matrix that is changed to a simple matrix nesting with `as.matrix`.

```
glmnet.coef<-as.matrix(coef(cv, s='lambda.min'))
```

The row names are appended as a separate vector along with column names. The `data.frame` function accepts the column name, words and then a vector of `row.names` from the prior matrix. The second column in the new data frame is named `glmnet_coefficients` and contains the numeric beta coefficients in the model.

```
glmnet.coef<-data.frame(words= row.names(glmnet.coef),
glmnet_coefficients=glmnet.coef[,1])
```

Reordering the data frame by decreasing coefficient value is done using `order` along with the `decreasing=T` parameter. This is done within the data frame using square brackets as if you were indexing specific values. Also, the vector of row names is changed to a categorical factor so `ggplot` can handle the terms more appropriately.

```
glmnet.coef<-glmnet.coef[order(
  glmnet.coef$glmnet_coefficients, decreasing=T),]
glmnet.coef$words<-factor(glmnet.coef$words,
levels=unique(glmnet.coef$words))
```

The `glmnet.coef` data frame contains a value for over 5000 terms along with a y intercept. Exploring the coefficients can be done by calling summary on the numeric vector. The lasso regression forced many of the term coefficients to zero which is evident in the `summary` results. The first and third quartile are 0.00 and the mean is 0.0013.

```
summary(glmnet.coef$glmnet_coefficients)
```

Using `subset` along with `length` you identify the number of words that have a positive and negative impact to being clickbait. Also, you visually see the coefficients being centered on zero, using a kernel density plot. As you mix the alpha parameter when training the model, more words will have non-zero coefficients, so the number of positive and negative words, and the kernel density plot, will change. The kernel density plot is represented in Figure 6.8.

```
length(subset(glmnet.coef$glmnet_coefficients,
glmnet.coef$glmnet_coefficients>0))
```

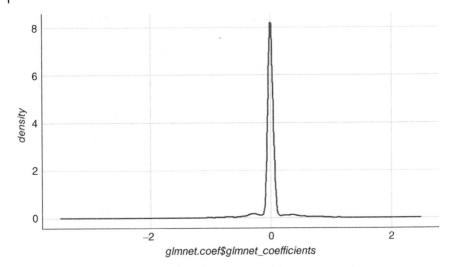

Figure 6.8 The kernel density plot for word coefficients.

```
length(subset(glmnet.coef$glmnet_coefficients,
glmnet.coef$glmnet_coefficients<0))
ggplot(glmnet.coef,
    aes(x=glmnet.coef$glmnet_coefficients)) +
    geom_line(stat='density', color='darkred',
      size=1) + theme_gdocs()
```

Of course, reviewing coefficient distributions is not that insightful, but it helps illustrate how a lasso regression differs from a typical linear regression model. It is more impactful to identify the specific terms at the distribution extremes. The individual coefficients represent the impact of the term on being clickbait when all other inputs are held constant. Thus reviewing single coefficients can be insightful for finding the most clickbait or legitimate headlines.

Using the head and tail functions along with an integer will identify the terms with the highest and lowest coefficients. This is because the glmnet.coef data frame was ordered previously. In this example, the new data frame top.coef has 20 rows. The data frame consists of the top and bottom ten terms according to the lasso regression coefficients. In the second line a new vector named impact is created. Using an ifelse statement, each row is labeled "Positive" or "Negative" representing the term's relationship to a headline being clickbait.

```
top.coef<-rbind(head(glmnet.coef,10),
tail(glmnet.coef,10))
top.coef$impact<-ifelse(
top.coef$glmnet_coefficients>0,"Positive","Negative")
```

A line segment plot can be constructed using the top.coef data frame. Using ggplot, pass in the data frame then specify x and y aesthetics. The first layer creates the line segments starting at the coefficient values and ending at 0. Each segment is arranged vertically by word along the y axis. The next layer adds a point at the beginning of each line segment. Each point is then color-coded according to the impact "positive" or "negative." Lastly theme_ few is applied to reduce visual clutter such as plot background. Changing the number of rows when constructing top.coef will change the number of terms in the visual. In this example, the line segment plot is exemplified in Figure 6.9.

```
ggplot(top.coef, aes(x=glmnet_coefficients, y=words)) +
  geom_segment(aes(yend=words), xend=0,
colour="grey50") +
  geom_point(size=3, aes(colour=impact)) + theme_few()
```

A binomial regression calculates a logit not a probability. As a result, the coefficients themselves do not have much meaning. For example, having a –2 logit coefficient does not translate to –2 "clickbaits." To examine an individual term's impact for increasing or decreasing the probability of being clickbait, the coefficient value is transformed using the "inverse logit" function. This converts a continuous value to a range between 0 and 1. The input is the logit value and the outcome is the more meaningful probability.

$$Inverse\,Logit\,Function = \frac{e^x}{\left(1+e^x\right)}$$

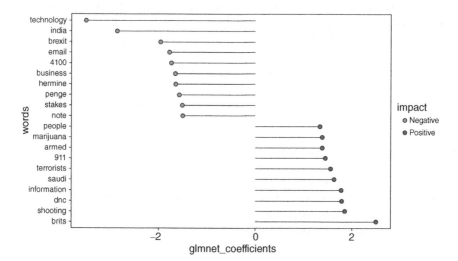

Figure 6.9 Top and bottom terms impacting clickbait classifications.

The `arm` library has a function called `invlogit`. The code below uses the `invlogit` function to transform logit coefficients to a probability between 0 and 1, while simultaneously adding a new vector. For comparison, this is applied to all coefficients using `glmnet.coef` and `top.coef` data frames.

```
glmnet.coef$probability<-invlogit(
            glmnet.coef$glmnet_coefficients)
top.coef$probability<-invlogit(
            top.coef$glmnet_coefficients)
```

An experiment outcome probability is never negative. In this data, 50% of the headlines were clickbait, so the training and test partition will be evenly split, or close to it. The model learned a natural state where half of all observations are clickbait. For practical purposes, probabilities below 50% can be thought of as reducing the likelihood that a headline is clickbait. Examining the `top.coef` data frame illustrates this point. All of the words labeled as negative according to beta sign have less than 0.50 probability but still higher than 0. The complete `top.coef` data frame is represented in Table 6.3, due to partitioning your results may vary.

Reviewing Table 6.3, having "brits" in a headline is a strong indication that the headline is clickbait. Conversely, a headline containing "technology" only increases the probability by 0.02.

To get a sense of the lasso effect on probability and where the `top.coef` terms lie among all word probabilities, construct a scatter plot. Using the base `plot` function the x values contain all word probabilities from `glmnet.coef$probability`. The y axis is captured as individual terms. Term probabilities are shown as circles with blue outlines. Next, add additional points from the `top.coef$probability` data. These have a different shape and are colored red. The final illustration is captured in Figure 6.10.

```
plot(glmnet.coef$probability,glmnet.coef$word,
col='blue')
points(top.coef$probability,top.coef$word, col='red',
pch=16)
```

In the Figure the vast majority of words have 0.50 probability. The lasso regression penalized words resulting in 0.00 coefficients. The penalized terms have no probability impact on the outcome because the natural state is half clickbait and half legitimate. Also, the outlier solid dots infer that the `top.coef` data frame is capturing enough words to be insightful. This figure can help you choose the appropriate number of top and bottom terms when constructing `top.coef`.

Table 6.3 The complete `top.coef` data, illustrating the negative words having a positive probability.

words	glmnet_coefficients	impact	probability
brits	2.49724	Positive	0.923948
shooting	1.847323	Positive	0.863812
dnc	1.788934	Positive	0.856797
information	1.776029	Positive	0.855206
saudi	1.628565	Positive	0.835973
terrorists	1.565807	Positive	0.827185
911	1.450905	Positive	0.810138
armed	1.385272	Positive	0.799836
marijuana	1.384973	Positive	0.799788
people	1.346696	Positive	0.793589
note	−1.50314	Negative	0.181957
stakes	−1.51377	Negative	0.180381
penge	−1.57527	Negative	0.171466
hermine	−1.6493	Negative	0.161203
business	−1.66078	Negative	0.159658
4100	−1.7356	Negative	0.149873
email	−1.77912	Negative	0.144412
brexit	−1.96833	Negative	0.122568
india	−2.86692	Negative	0.053813
technology	−3.50817	Negative	0.029081

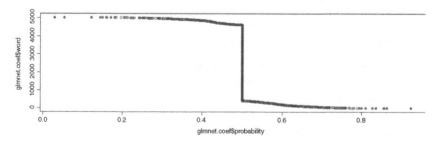

Figure 6.10 The probability scatter plot for the lasso regression.

6.2.6   Case Study Wrap Up: Model Accuracy and Improving Performance Recommendations

In the final step of the text mining workflow you revisit the original problem statement.

7) **Reach an insight or recommendation.** You will know the top terms signaling clickbait. Further, you will be able to recommend ways to improve the algorithm's accuracy.

The lasso regression model performed well with a small number of headlines. A test set accuracy above 0.70 shows that the method has merit. As a text miner, you should feel comfortable that the method has sound performance but that it must be improved before being implemented. Possible methods to improve the model include:

- **Feature engineering** – create additional inputs for the number of characters or words per headline. Another piece of information could be a timestamp or the source. (For educational purposes, source was omitted as an input because the data was limited.)
- **Tokenization** – experiment with bi-gram and tri-gram tokenization as model inputs. For example, the word "good" may have a beta coefficient but negation words ahead of "good" would change the coefficient sign completely. Using bi-grams would capture an input "not good" that may have more explanatory power than "good."
- **Alpha penalty** – adjust the penalty parameter to balance the number of inputs and accuracy. If a model has more inputs, it will take longer to make a prediction. When implemented, at scale this can have a negative impact to user experience.
- **Prediction stacking and ensemble modeling** – combine more models using different methods such as lasso and naïve Bayes into a final more complex model. Machine learning books explain how to stack and ensemble so it is not covered here.
- **More data** – increase the number of headlines and sources. Adding more data increases modeling building time but will likely significantly improve the results. This case study utilized 3000 headlines but model fitting using tens of thousands of headlines improves the information diversity.

In addition to these technical improvements you can confidently illustrate the impact of specific words. Simply flagging headlines with the top words in Table 6.3 could be a short-term solution to classifying clickbait in a non-production environment. However, clickbait profiteers continually adjust tactics, and headlines change. As a result, you must continually retrain the model with current headlines.

6.3 Summary

In this chapter you learned a basic document classification approach. Document classification algorithms are in use every day to help protect against email spam and social media "clickbait," but there are more applications including journalism and document archiving. Specifically you learned:

- how to create a DTM from a training set and use it when constructing new DTMs for new documents
- what a sparse model matrix is
- how to create a data partition for training and testing of the document classification algorithm
- how to create a document classification algorithm
- what lasso regression is and how it differs from linear regression
- perform cross validated lasso regression
- how to calculate the area under the curve (AUC) for the document classifier
- how to calculate the overall accuracy of the classifier on unseen headlines
- how to identify the words most contributing to clickbait headlines.

7

Predictive Modeling: Using Text for Classifying and Predicting Outcomes

In this chapter, you'll learn

- how classification is different from continuous prediction
- how to create a DTM using `text2vec` instead of the `tm` package
- how to apply the training set vocabulary to new text
- how to create a classification algorithm for hospital readmissions, based on discharge text and patient information
- how to create a continuous predictive model of social media shares based on article text.

7.1   Classification vs Prediction

Chapter 7 extends the document classification approach of Chapter 6 in new ways. First, you will learn to apply an elastic net linear model for patient outcome classes. Next the same algorithm will be fitted to movie reviews and revenue. In the previous chapter, the elastic net (more specifically lasso) regression was applied to classify new document types. However, the outcomes in Chapter 7 are not document types. Instead, the text is an input to an external, non-document outcome.

There are many books devoted to machine learning. This book touches on basic principles of machine learning in various sections. It should be noted that the topic of machine learning is large and varied beyond this chapter's description. As you build your text mining skills, you will add new machine learning algorithms and practical use cases to your tool set. Chapters 6 and 7 are meant to give you a foundation to build upon. The elastic net algorithm was chosen for its simplicity and general applicability, but other methods and preprocessing steps can improve accuracy for your own text mining models.

In this chapter, you will employ both classification and continuous prediction examples using text as inputs. A simple way to understand the difference between classification and continuous prediction is that classification answers a question

Text Mining in Practice with R, First Edition. Ted Kwartler.

from among specific choices. Asking "Will you finish this chapter?" requires a "yes" or "no" answer. Similarly, "Will a customer buy product A, B or C?" can be answered with one of the three choices. Asking questions with defined "class" outcomes represents classification. In contrast, prediction asks a different style of question. For example, "How many pages will you read?" requires a numeric answer. More specifically, the outcome, pages, is continuous. You could read 0 to some practical page limit. Another predictive outcome includes predicting how much a customer will spend. This contrasts the previous A, B, or C example, because the dollars range from 0 to some limit of their income or credit.

Understanding the differences between classification and prediction is paramount to learning supervised learning. Training data and algorithm approaches are adjusted based on the classification or predictive outcome itself. If you choose an approach incorrectly, the outcome will be useless. For example, consider an algorithm answering "Will Cleveland or Golden State win the basketball game?" This is a binary outcome with one of two teams being most likely to win represented as 0 and 1. If you incorrectly choose a continuous predictive algorithm, the output could be larger than 1! This makes no sense, because a team cannot have a win probability greater than 1 or win the same game twice. The inverse is also true. If you were predicting the number of points that the Cleveland team was going to score in a game, choosing a classification algorithm would not make sense. The outcome of a classification algorithm would be a probability. A team needs to score more than 0 to 1 points to win.

The distinction between classification and prediction is so important that this chapter is unlike the previous ones. This chapter contains *two* mini case studies. The aim is to let you execute and learn both outcome types. Along the way, the chapter's code snippets can be compared and contrasted. By using code for both classification and prediction, you can identify the correct methods when modeling your own text mining projects.

7.2 Case Study I: Will This Patient Come Back to the Hospital?

Hospitals are rightfully concerned about patient readmissions. According to the Agency for Healthcare Research and Quality, over 11 months in 2011, hospitals in the US spent $41 billion dollars treating patients within 30 days of an initial discharge. Further, both Medicare and Medicaid services fine hospitals if readmission rates are too high. From a patient advocacy perspective, hospital readmissions cause undue hardship and strain too. Thus, it is in everyone's best interest to identify patients at the moment of discharge who are likely to come back within 30 days. Once likely readmission patients are identified, hospitals can review cases to understand the reasons behind readmission or, more practically, administrators can keep the most likely patients in the hospital for additional care, avoiding fines and extra costs and helping to the patients recover more fully.

In this example, you will build a model using patient data from 8500 diabetic patients. The data was originally part of an academic study concerning readmission. The data has been organized and cleaned so that models can be built quickly, leaving you to learn the underlying text mining fundamentals.

7.2.1 Patient Readmission in the Text Mining Workflow

Although the workflow outlined in the book is specific to text mining projects, it can be coerced into a traditional machine learning workflow.

1) **Define the problem and specific goals.** Compare diabetic patient readmission models with and without discharge notes as inputs. Apply the best model to newly discharged patients to identify diabetic patients that are likely to revisit the hospital within 30 days.
2) **Identify the text that needs to be collected.** A sample of 8500 patient records has been collected, including discharge notes. This data represents the training and validation data for model building.

Getting the Data

Please navigate to www.tedkwartler.com and follow the download link. For this analysis, please download "diabetes_subset_8500.csv." The file is a sample of 8500 diabetic patient records from an academic study. (https://www.hindawi. com/journals/bmri/2014/781670/cta/) The file contains 136 columns including gender, race, patient weight and discharge information. The study's raw inputs have been preprocessed as dummy variables and cleaned for missing values.

---

3) **Organize the text.** The discharge notes will be organized into a DTM. The matrix will be used for modeling and then will be combined with the other patient data to fit a new model for comparison.
4) **Extract Features.** Elastic net models will be created to calculate a patient's readmission likelihood.
5) **Analyze.** Modeling metrics with and without text inputs will be compared. The best performing model will be selected.
6) **Reach an insight or recommendation.** Apply the best model to a new patient record, recommending if the new patient is a candidate for readmission.

7.2.2 Session and Data Set-Up

In the last chapter, the vocabulary matching approach followed the RTextTools method. In Chapter 6, you either corrected a typo in the `create_matrix` function or applied a custom `match.matrix` function that was provided. In this chapter, you will use the `text2vec` package to create the vocabulary and training DTM. The vocabulary is then applied to construct a train set DTM and also the new patient data. You should be familiar with `text2vec` because a previous chapter covered vector distance measures.

The `text2vec`'s text organization functions are illustrated to show another text organization method that can be used for machine learning. Next `caret` is used for data preparation. The `tm` package is added for preprocessing functions. Once again the `glmnet` library is loaded to fit the elastic net regression. Finally, the `pROC` library helps to visualize model performance.

```
library(text2vec)
library(caret)
library(tm)
library(glmnet)
library(pROC)
```

The code below is another example of a custom preprocessing function. Throughout the book, similar functions help reduce errors and make the code more concise. Here only two functions are applied but you can add more if needed.

```
diagnosis.clean<-function(x){
  x<-removePunctuation(x)
  x<-stripWhitespace(x)
  return(x)
}
```

After reading in the patient data and reviewing you may notice the text columns are held in `diag_1_desc`, `diag_2_desc` and `diag_3_desc`. Each of these columns are actually high cardinality factors not truly natural language text. For instance, `diag_1_desc` contains 450 diagnosis description levels among the 8500 patients. Rather than create hundreds of dummy variables for each column another approach is to treat the many factor levels as text. To do so, a new data feature called `diag.text` is added to the diabetes object. The three columns are combined using `paste` and then `as.character` converts the columns to strings instead of factors.

```
diabetes<-read.csv('diabetes_subset_8500.csv')

diabetes$diag.text<-
as.character(paste(diabetes$diag_1_desc,
  diabetes$diag_2_desc, diabetes$diag_3_desc, sep=' '))
```

The custom function `diagnosis clean` is applied to the final text vector.

```
diabetes$diag.text<-diagnosis.clean(
      diabetes$diag.text)
```

Using caret's `createDataPartition` 70% of the records are assigned to train with the remaining patient rows becoming the test set.

```
train<-createDataPartition(diabetes$readmitted,p=.7,
list=F)
train.diabetes<-diabetes[train,]
test.diabetes<-diabetes[-train,]
```

Up to this point, the setup has been consistent with other chapters, but the `text2vec` library differs in DTM construction because of the use of an iterator. An iterator is a method of applying functions *in* an object. The function is applied as the iterator winds its way through the object rather than to an entire object. This means that an iterator can apply functions to large objects that may be too large to fit in memory. Here `itoken` iterates through `diag.text`. Along the way `tolower` is applied along with another function `word_tokenizer`. The `word_tokenizer` function wraps `str_split` to separate individual words.

```
iter.maker<-itoken(train.diabetes$diag.text,
        preprocess_function = tolower, tokenizer =
word_tokenizer)
```

The iterator object is a set of instructions that are passed to another function `create_vocabulary`. Here `v` is a list of unique words and statistics from the diagnosis text.

```
v <-
create_vocabulary(iter.maker,stopwords=stopwords('en'))
```

Next, the vocabulary object is passed to a "vectorizer." The vectorizer ultimately lets you create a corpus object by instructing R to make a vector of terms.

```
vectorizer <- vocab_vectorizer(v)
```

The `vectorizer` is needed to construct a DTM using the `text2vec` package. Another iterator is needed to create the DTM so `itoken` is used again. The `it` object along with the original `vectorizer` is passed to the `create_dtm` function to get a matrix.

```
it <- itoken(train.diabetes$diag.text,
        preprocess_function = tolower,
        tokenizer = word_tokenizer)
dtm <- create_dtm(it, vectorizer)
```

This may seem convoluted, so let's deconstruct the steps. An iterator is sometimes needed to learn statistics about text that is too large for in-memory analysis. So an iterator containing a set of instructions is created solely to explore the text's vocabulary. The vocabulary information is held in a list but must be changed to a vector in order to construct a DTM. So the iterator is "vectorized" using `vocab_vectorizer`. The first iterator is only used for vocabulary construction. Then another iterator is needed to wind its way through the text again. The second iterator is passed to `create_dtm` with the vectorized vocabulary from the first iterator's analysis. The confusing part is the need for two iterators, but they are used for different purposes. The first is for vocabulary and the second for the matrix construction.

The `dtm` object is based on simple term frequency. When working on other data you may need TF-IDF weighting instead. The code below is not needed for the case study but demonstrates TF-IDF weighting using `text2vec`. First extract the inverse document frequency information using `get_idf`. Next transform the `dtm` using `transform_tfidf` to change the DTM values.

```
idf<-get_idf(dtm)
dtm.tfidf<-transform_tfidf(dtm,idf)
```

7.2.3 Patient Modeling

The `dtm` can be used to fit an elastic net binomial regression. The `cv.glmnet` function needs inputs, then the outcome variable along with tuning and other parameters. In this case, the model inputs are solely constructed from the diagnosis text. The outcome variable, "readmitted", contains two classes. As a result "binomial" must be specified as the family. If the outcome had more than two class levels then the family would be "multinomial." In this example, the alpha mixing parameter is 0.9 and the accuracy measure is AUC. This is consistent with the ROC curve, but you could specify a different accuracy measurement. As a cross validated model the number of folds must be specified. Finally, the y intercept is forced to zero because `intercept=F`.

```
text.cv<-cv.glmnet(dtm,y=as.factor(
train.diabetes$readmitted), alpha=0.9,family='binomial',
type.measure='auc', nfolds=5, intercept=F)
```

As shown in Chapter 6, the model performance can be visually inspected using `plot`. Figure 7.1 shows the AUC information for `text.cv`. Review the plot explanation in Chapter 6 to interpret this plot.

```
plot(text.cv)
```

In many cases, numeric and dummy inputs are more accurate than text inputs. It is a best practice to compare models made of non-text attributes. In this case, the non-text patient information can be selected using the code below. The training data is indexed to include all rows, but only the first 132 columns, thereby dropping all text columns.

```
no.text<-as.matrix(train.diabetes[,1:132])
```

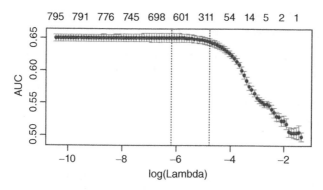

Figure 7.1 The text-only GLMNet model accuracy results.

The no.text matrix is once again passed to the cv.glmnet function. The dependent variable and other input parameters remain the same.

```
no.text.cv<-cv.glmnet(no.text, y=
as.factor(train.diabetes$readmitted), alpha=0.9,
family='binomial',type.measure='auc', nfolds=5,
intercept=F)
```

The no.text.cv model has a slightly improved the AUC. Reviewing Figure 7.2, the AUC is above 0.75 on the training data.

```
plot(no.text.cv)
title("GLMNET No Text")
```

The results are not entirely surprising. The text inputs contain limited information from three columns while the non-text columns have medical history and patient demographics. The non-text inputs simply contain more information for

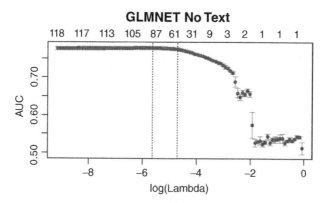

Figure 7.2 The numeric and dummy patient information has an improved AUC compared to the text only model.

the model to learn from. However, a data scientist can often squeeze a bit more accuracy out of a model by combining numeric and text inputs into a model.

Use cBind to combine the no.text and dtm matrices. Recall that the dtm is a dgcMatrix which is a sparse data object from the Matrix package. The base cbind function will not work on a dcgMatrix so use the cBind version instead. The resulting all.data sparse matrix has 5950 patient rows with 1173 columns.

```
all.data<-cBind(dtm,no.text)
```

The new matrix is once again passed to cv.glmnet. The input parameters are consistent with the previous two models.

```
all.cv<-cv.glmnet(all.data,y=as.factor(
train.diabetes$readmitted), alpha=0.9, family='binomial',
type.measure='auc', nfolds=5, intercept=F)
```

The combined plot shows an improved AUC in Figure 7.3. More model key performance indicators (KPIs) are explored in the next section, but Figure 7.3 does demonstrate that adding text helps improve results.

7.2.4 More Model KPIs: AUC, Recall, Precision and F1

Although you can visually tell the difference, the three models need to be compared explicitly. Following R's model predict function pass in the model, text.cv, then the input matrix called dtm. The response type is "class" so that the outcome class is returned instead of each patient's exact probability. The last parameter specifies a lambda value that minimizes the error instead of the most parsimonious one. The predict function will return each patient's class, True or False, as a string. The class values need to be changed from

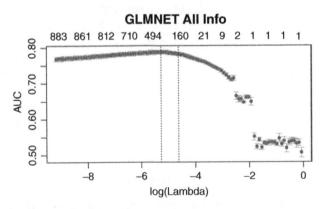

Figure 7.3 The cross validated model AUC improves close to 0.8 with all inputs.

strings to the logical classes strictly representing True or False. As a result, the class predictions are nested in as.logical.

```
text.preds<-as.logical(predict(text.cv,
  dtm,type='class', s=text.cv$lambda.min))
```

Next let's calculate the ROC and plot it. Using the pROC library's roc function, pass in both the actual patient outcomes and the model predictions. Each has to be multiplied by 1 to change the factors to 1 or 0.

```
text.roc<-roc((train.diabetes$readmitted*1),
text.preds*1)
```

In this case, the text-only model has an AUC of 0.637. Now the same code can be reused for the remaining two models. Remember that the no.text.cv model was fit using only the numeric patient data. The last model all.cv was created with all information numeric and text and held in the all.data matrix.

```
no.text.preds<-as.logical(predict(no.text.cv,
no.text,type='class', s=no.text.cv$lambda.min))
```

```
no.text.roc<-roc((train.diabetes$readmitted*1),
no.text.preds*1)
```

```
all.data.preds<-as.logical(predict(all.cv,
all.data,type='class', s=all.cv$lambda.min))
```

```
all.data.roc<-roc((train.diabetes$readmitted*1),
all.data.preds*1)
```

The three AUC values are shown in Table 7.1. You should notice that the weakest model is the text-only inputs. The best model uses both numeric and text thereby adding a bit more accuracy including discharge notes. Considering the staggering costs of hospital readmissions, a model maximizing AUC and including text is worth the additional computational time.

Use the code below to visually compare the ROC curves. The first line creates the plot with the text model's information. The next two add a new line in a different color and with different styles for comparison. Figure 7.4 shows the ROC curves for the models and the additional lift above random for each of the models.

Table 7.1 The AUC values for the three classification models.

ROC object name	AUC
text.roc	0.637
no.text.roc	0.6741
all.data.roc	0.728

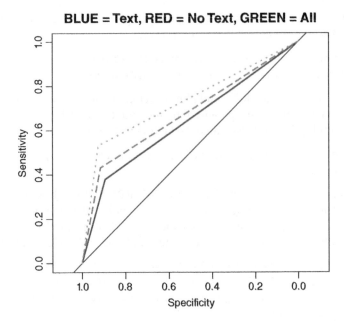

BLUE = Text, RED = No Text, GREEN = All

Figure 7.4 The additional lift provided using all available data instead of throwing out the text inputs.

```
plot(text.roc,col="blue",main="BLUE = Text, RED = No
Text, GREEN=All",adj=0)
plot(no.text.roc, add=TRUE,col="red", lty=2)
plot(all.data.roc,add=TRUE,col="darkgreen", lty=3)
```

7.2.4.1 Additional Evaluation Metrics

There are additional KPIs for binary classification models. Popular ones including recall, precision and the F1 score could have been calculated using the document classification model of Chapter 6. These KPIs were not covered in Chapter 6, so that you could focus on the underlying basics of the model and the introduction of some machine learning principles. Now that you have added sufficiently to your text mining skills it makes sense to cover recall, precision and F1 scores.

A great visual to understand recall and precision comes from Wikipedia and is shared in Figure 7.5, where you are presented with an entire population represented as dots. Each dot could represent a patient from the data set. In this case, the best classification model, `all.cv`, makes an assessment of each patient's outcome. Sometimes the model correctly identifies the true success class which is *not* being readmitted. The model can also identify the true readmission which represents an unsuccessful hospital discharge. Still other times

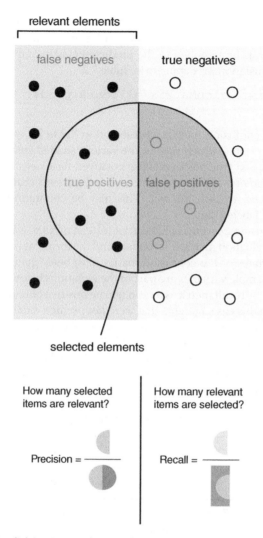

Figure 7.5 Wikipedia's intuitive explanation for precision and recall.

the model classifies patients incorrectly, meaning that some patients are pre-dicted not to readmit yet show up back at the hospital. Similarly, the model may predict that a group of patients will readmit but they go home never to be seen again (at the hospital).

In the case study, a False prediction is actually the success class because that means the patient did not come back to the hospital. Therefore the *true positives* are cases in which the model predicted False for readmission and the patient actu-ally did not readmit. The subsequent explanation will make this less confusing.

The four outcome states in the visual can be captured in a confusion matrix using the code below. Rather than use the `table` function, the `confusion-Matrix` function is first applied to the predicted classes and then the actual classes. The confusion matrix appears in Table 7.2.

```
confusion<-confusionMatrix(all.data.preds,
train.diabetes$readmitted)
```

Reviewing the confusion matrix represented in Table 7.2, you should note the four outcome states, which map to the various shaded areas in Figure 7.5. The states are labeled true positives, false positives, false negatives, true negatives. *Keep in mind that the true positive cases are patients that were predicted not to readmit and actually did not.* This may be counterintuitive because readmissions are labeled True.

The benefit of using `confusionMatrix` instead of `table` is that the function calculates evaluation metrics. To understand the formulas refer to Table 7.3, where the four states of the confusion matrix have been changed to example variables. The example variables are part of the equations below for reference.

The sensitivity or recall metric is a ratio of true positive outcomes to all positive outcomes, in this case, patients that were not predicted to readmit divided by all patients that actually did not readmit. Remember that the success class is not coming back to the hospital. Using Table 7.3 as a guide, the sensitivity formula is below. At 0.92, this model can be said to be highly sensitive because it is mostly accurate at identifying non-readmitted patients which is a successful outcome for the hospital.

Table 7.2 The best model's confusion matrix for the training patient data.

		Actual classes	
		F = not readmitted	T = readmitted
Predicted classes	F = not readmitted	3639	952
	T = readmitted	288	1071

Table 7.3 The confusion matrix with example variables.

		Actual classes	
		F = not readmitted	T = readmitted
Predicted classes	F = not readmitted	A = true positives	B = false negatives
	T = readmitted	C = false positives	D = true negatives

$$Sensitivity\ or\ Recall = \frac{A}{(A+C)}$$

The recall metric can be extracted from the confusion object using the following code snippet.

```
recall <- confusion$byClass['Sensitivity']
```

The next metric is called specificity or true negative rate. To calculate the metric, the true negative outcomes are divided by the sum of all negative outcomes. In this case, 1071 is divided by the sum of 952 and 1071. The model is less specific because the value is only 0.52. It is the probability of successfully being classified as readmitted, given all patient's readmissions.

$$Specificity\ or\ TNN = \frac{D}{(B+D)}$$

Another important metric is prevalence, which is a measure of how often non-readmission occurs in the population or the sum of all success classes. The formula is the sum of true success class divided by the entire sum of the table. In this example, the prevalence of non-readmission (hospital success) is 0.66. This is the probability of all patients for a successful outcome which is not coming back to the hospital.

$$Prevalence = \frac{(A+C)}{(A+B+C+D)}$$

The positive predicted value (PPV) is also known as the precision. This value is represented in the next formula. This model is relatively precise at identifying the successful patient outcomes with a 0.79 result.

$$\frac{PPV\ or}{Precision} = \frac{(Sensitivity * Prevelance)}{((Sensitivity * Prevelance) + (1 - Specificity) * (1 - Prevelance))}$$

To extract the precision, use the following code.

```
precision <- confusion$byClass['Pos Pred Value']
```

Another classification model performance indicator is called the F1 score, which attempts to unify the recall and precision into a single metric. There are variations of the F1 score providing different weights to precision or recall. This example is straightforward because precision and recall are equally weighted and is called the F1 score. The formula for F1 is two times the ratio of Precision times Recall divided by Precision plus Recall:

$$F1 = 2 * \frac{Precision * Recall}{Precision + Recall}$$

The model's F1 score is a respectable 0.85. It can be calculated using the R code here.

```
f1.score <- 2 * ((precision * recall) / (precision +
recall))
```

There are other more sophisticated metrics produced by the `confusion-Matrix` function. In practice the basic ones documented here serve as a good foundation to understand a model's effectiveness. When creating and choosing a machine learning model a good practice is to understand the effects of high precision with low recall or the opposite. Overall accuracy may not be the best success criterion, particularly with highly unbalanced classes, or when the outcome has significant monetary or human impact.

7.2.5 Apply the Model to New Patients

Thus far the `all.cv` model has been evaluated using the training set. Once you have explored tuning parameters such as adjusting alpha, the model should be applied to test data to ensure consistency. To do so the terms of the DTM from the training data must match the test set DTM. The `text2vec` library conveniently employs another iterator, the original `vectorizer` object and a function called `create_dtm` to recreate the DTM columns. This means that the `match.matrix` function from Chapter 6 is not necessary.

The `test.it` iterator is applied to the test set text along with a preprocessing function and the standard `word_tokenizer`. The `test.it` object is then passed to the `create_dtm` function. The second parameter of the `create_dtm` function must refer to the original training set word vector so that the DTM terms match. If your training DTM was TF-IDF weighted you would apply the appropriate code to the DTM to ensure that the training and test DTMs were weighted similarly. Although no TFIDF weighting is applied in this example, the code was previously illustrated.

```
test.it<-itoken(test.diabetes$diag.text,
          preprocess_function = tolower,
          tokenizer = word_tokenizer)
test.dtm<-create_dtm(test.it,vectorizer)
```

The test set DTM now has the same number of columns as the training set DTM. The `test.dtm` can now be appended to the numeric test set data. As before, the individual text columns are dropped, only leaving the numeric data. Then the numeric test set data and `test.dtm` are bound using cBind.

```
test.no.text<-as.matrix(test.diabetes[,1:132])
```

```
new.patients<-cBind(test.dtm,test.no.text)
```

With a similar model matrix, the predictions can be made on the new. patients. Once again, predict is passed the all.cv model then the model matrix. The response type is specified as class and the s value represents the lambda value that minimizes the error.

```
test.preds<-predict(all.cv,new.patients,
type='class',s=all.cv$lambda.min)
```

A quick assessment of the confusion matrix shows that this model is behaving similarly compared to the training evaluation metrics. The code to create the test set confusion matrix and F1 score is below, and the test confusion matrix is represented in Table 7.4.

```
test.confusion<-confusionMatrix(test.preds,
test.diabetes$readmitted)
test.precision <- test.confusion$byClass['Pos Pred
Value']
test.recall <- test.confusion$byClass['Sensitivity']
test.f1 <- 2 * ((test.precision * test.recall) /
(test.precision + test.recall))
```

Overall, the model is performing consistently with unseen data. Assuming that the defined problem or ecosystem does not change, the model will behave as expected once implemented. The training set Precision was 0.79 compared to the test Precision 0.77. Recall is consistent, moving from 0.92 on training to 0.91 for the test data. Likewise, the F1 score is stable moving from 0.85 to 0.83. Depending on the use case, the model metrics may be acceptable but in significant outcomes, such as life and death implications, you may need to add more data or information to improve the model's performance.

7.2.6 Patient Readmission Conclusion

Patient readmission is a costly problem in the US. Hospital administrators and academics review this issue in the hopes of improving financial and humanistic

Table 7.4 The test set confusion matrix.

		Actual classes	
		0 = not readmitted	1 = readmitted
Predicted classes	0 = not readmitted	1543	454
	1 = readmitted	140	413

outcomes. In this example, you created a simple patient readmission model specific to diabetic patients. The data was collected from an academic study but was cleaned so you could focus on learning the `text2vec` method of DTM construction for both train and test cases. With so much at stake, practitioners apply additional sophistication to improve accuracy. As you explore and tweak the code, keep in mind that you can use other algorithms, stack text predictions as an input to other information or ensemble models to improve accuracy.

In this example, text mining for classification outcomes could have a constructive impact on society. I encourage you to download the complete and raw data set at www.tedkwartler.com and explore ways to improve results. Along the way you may add to your machine learning toolkit and could possibly positively affect others.

7.3 Case Study II: Predicting Box Office Success

In this exercise, you will predict a movie's opening weekend revenue based on early reviews. Once a movie is created, early screenings are shared with prominent news outlets. The hope is that a positive review will result in additional "buzz" and more ticket sales when the movie opens publicly. At this point, the movie has been created, so the production costs are sunk. However, the reviews may indicate a surprising success requiring additional marketing dollars and wider release to additional theaters. In contrast, poor reviews may indicate that a movie could flop, meaning that the studio should reduce marketing spend and overall release. Your model's inputs will be early screening reviews but in a real application additional inputs such as holiday releases and specific actors or directors are used to improve results. In both cases, the dependent variable is a continuous dollar amount from revenue on the opening weekend.

An interesting exercise, rumored to be used at Netflix, would be to text mine movie scripts for attributes and story arcs to predict viewership on the platform. As you perfect your predictive text mining capabilities remember that the most prized model would be one that predicted commercial success from a script, with expected directors, likely actors, and expected release dates prior to costly production.

Of course, other predictive text mining uses exist. Entire services are set up to understand and interpret text in the financial sector and elsewhere. For example www.stocktwits.com is a Twitter-like service specifically used for financial and stock level dialogue. Using this service, people have collected the information and attempted to predict stock performance. In a related context, others have modeled daily bitcoin prices based on bitcoin forum dialogue. The following movie reviews example may seem unusual but is illustrative of the techniques needed in more mainstream and impactful situations.

7.3.1 Opening Weekend Revenue in the Text Mining Workflow

Although the workflow outlined in the book is specific to text mining projects it can be coerced into a traditional machine learning workflow.

1) **Define the problem and specific goals.** Predict a movie's opening weekend revenue using a machine learning model.
2) **Identify the text that needs to be collected.**

Getting the Data

Please navigate to www.tedkwartler.com and follow the download link. For this analysis, please download "2k_movie_reviews.csv." The file is a sample of 2000 movie reviews collected from a 2010 academic study. (http://www.cs.cmu. edu/~ark/movie$-data/) The file contains three columns including the movie review, the original XML file indicating the film name and the opening weekend results.

---

3) **Organize the text.** Using the `text2vec` iterator you will create a training data DTM and a holdout test set DTM.
4) **Extract Features.** An elastic net model will be created to predict the continuous outcome `opening_weekend` variable.
5) **Analyze.** Modeling metrics, including MSE and MAE, will be examined for the training set.
6) **Reach an insight or recommendation.** Using the model recommend `opening_weekend` revenue for holdout movie reviews.

7.3.2 Session and Data Set-Up

Much of this script repeats the previous classification example. This is because the basic machine learning steps of partitioning and DTM construction are the same. Load the specific libraries into your R session. By now you are familiar with most of the packages. The `Metrics` package provides functions for evaluating model performance. For the classification example the AUC was calculated using pROC. Alternatively, the `Metrics` package also contains the same AUC calculation. In contrast, the `Metrics` package has evaluation metrics used for continuous outcomes.

The rest of the packages include `data.table`, `pbapply` and `text2vec` for data organization. The `caret` and `glmnet` libraries are used in data partitioning and model building. The text cleaning functions from `qdap` and `tm` are once again used. Lastly the `Metrics`, `tidyr` and `ggthemes` packages are used for output evaluation.

```
library(data.table)
library(pbapply)
library(text2vec)
library(caret)
library(glmnet)
library(qdap)
library(tm)
library(Metrics)
library(tidyr)
library(ggthemes)
```

To get started create move.data with fread on the "2k_movie_reviews.csv" file. The movie.data object contains three attributes from 2000 movies.

```
movie.data<-fread('2k_movie_reviews.csv')
```

Before constructing a DTM, create a custom cleaning function. The text will be passed into the review.clean function. Typical cleaning functions such as removePunctuation, stripWhitespace, removeNumbers and tolower are applied. Less common but still helpful for term aggregation include replace_contraction and stemmer from qdap. The stemmer is really a wrapper of stemDocument from tm and will perform the same operations. The functions are applied in order to clean the text and return it.

```
review.clean<-function(x){
  x<-replace_contraction(x)
  x<-removePunctuation(x)
  x<-stripWhitespace(x)
  x<-removeNumbers(x)
  x<-tolower(x)
  x<-stemmer(x)
  return(x)
}
```

The object clean.text represents a character vector with the cleaning functions applied. In this case, 2000 reviews are cleaned in turn which can take some time, depending on your computer's RAM. Notice that no stopwords have been removed. This step will be applied later as part of a text2vec function.

```
clean.text<-review.clean(movie.data$train.movies)
```

Creating a separate object representing the dependent variable helps to keep the overall data processing clean and separated from the raw data as it was read in. Here y is created from the opening_weekend variable.

```
y<-movie.data$opening_weekend
```

Next, the clean data and dependent variable need to be partitioned. Although the elastic net model will be cross validated, the small number of reviews means it is a good idea to have a holdout. Using `createDataPartition`, 80% of the y variable is selected. Partitioning in this manner is more sophisticated than a simple random sample. If your y variable is categorical the function tries to balance the class distributions between the training and validation sets. In this case, the outcome is numeric so the sample is split into grouped sections based on percentiles, and sampling is done within these subgroups. In both cases, the `createDataParition` function works to ensure that the test set mimics the training set. In contrast, a random sampling could yield a split that does not reflect all observations.

```
train<-createDataPartition(y,p=0.8,list=F)
train.movies<-clean.text[train]
train.y<-y[train]
test.movies<-clean.text[-train]
test.y<-y[-train]
```

You will create an iterator similar to the classification example earlier in this chapter. Once again, `iter.maker` is constructed using `itoken`. It is then passed to the `create_vocabulary` function. However, this example deviates slightly by employing a new stopwords list. The "SMART" stopwords list was constructed from Massachusetts Institute of Technology (MIT) research. The MIT list contains more words than the English list. The terms "movie" and "movies" are concatenated to this larger stopwords list. The end result is a vocabulary with terms, term counts and document counts.

```
iter.maker<-itoken(train.movies, tokenizer =
word_tokenizer)
v <- create_vocabulary(iter.maker,
stopwords=c(stopwords('SMART'),'movie','movies'))
```

Compared to the classification example, the code below represents a new step. The original vocabulary from 2000 movie reviews contains more than 43,000 unique terms after removing stopwords! Recall that we previously used `removeSparseWords` from tm to reduce the terms because the DTM is extremely sparse. The text2vec package provides `prune_vocabulary` for this purpose. Not only does `prune_vocabulary` remove infrequent terms, but the function has the added benefit of also removing very frequent terms not accounted for in the stopwords that likely yield little explanatory power. This function accepts the vocabulary and parameters needed to reduce the number of terms. Using `term_count`, an individual word must appear in at least 10 documents. The terms are further limited because an individual term cannot be in more than half of all documents because of the `doc_pro-portion_max` parameter. These would represent a common and non-informative term. The last parameter, `doc_proportion_min`, further limits the vocabulary. With 1602 documents in the training set a term must appear in 1.6 (= 0.001*1602) or more reviews to be included – essentially two or more reviews. This means a term that appears in a single review would be excluded altogether. This parameter is used to exclude outlier terms in a vocabulary.

```
pruned.v<-prune_vocabulary(v, term_count_min = 10,
doc_proportion_max = 0.5, doc_proportion_min = 0.001)
```

The smaller vocabulary of 7427 terms will decrease the modeling construction time. The reduced time stems from the fact that the elastic net algorithm has fewer document attributes to assess. The new pruned vocabulary is passed to the `vocab_vectorizer` similarly to the classification example. Then another iterator is constructed with the `train.movie` cleaned text. The `it` object is then passed to `create_dtm` along with the `vectorizer` referencing the pruned 7427 terms. The final object, `dtm`, has 1602 rows corresponding to each review in the training set. The DTM's 7427 columns represent the unique terms found among all reviews in the training set.

```
vectorizer <- vocab_vectorizer(pruned.v)
it <- itoken(train.movies, tokenizer = word_tokenizer)
dtm <- create_dtm(it, vectorizer)
```

7.3.3 Opening Weekend Modeling

You can now fit a cross validated model using `cv.glmnet` with the `dtm`. You have used this function previously but the `family`, `measure` and `intercept` parameters have changed. A "Gaussian" `family` ensures that the outcome is continuous. This contrasts with document classification and the patient readmission model, where the outcome was binary. Using `intercept=T` further instructs the function to allow an intercept value rather than force the linear starting point, β_0, to be zero. This means that the y intercept will not be the origin when an x,y plot is constructed. Lastly, `measure` is now "mse" signifying "mean squared error."

> The "mse" is an average of all squared errors. To understand "mse," perform all operations in reverse. First find the error by calculating the difference between an actual movie's revenue and the predicted revenue. In this model an error represents the dollar difference between a movie's predicted and actual opening weekend revenue. Next, square the error representing the predicted revenue missed. This is squared so that positive and negative errors do not cancel each other out in the last step. Lastly an average is applied to all squared error values. For example, consider three fictitious movies in Table 7.5.

Table 7.5 Three actual and predicted values for example movies.

Actual revenue	Predicted revenue
$2,000,000	$2,100,000
$1,000,000	$900,000
$1,250,000	$1,250,000

Table 7.6 With error values calculated.

Actual revenue	Predicted revenue	Error
$2,000,000	$2,100,000	−$100,000
$1,000,000	$900,000	$100,000
$1,250,000	$1,250,000	$0

Table 7.7 With squared error terms.

Actual revenue	Predicted revenue	Error	Squared errors
$2,000,000	$2,100,000	−$100,000	$\$^2$10,000,000,000
$1,000,000	$900,000	$100,000	$\$^2$10,000,000,000
$1,250,000	$1,250,000	$0	$\$^2$0

To calculate an error, subtract the actual values from the predicted values. In Table 7.6 a new column labeled "Error" has been added.

Without squaring the errors, the first two movie error values cancel each other out. An average of the error terms is 0 ((−$100k + $100k +0) / 3). The numerator becomes 0 divided by 3. This example model was not perfect, so taking a straight average at this point is misleading. The error terms have to be squared in Table 7.7.

Now, a simple mean of $\$^2$10,000,000 and $\$^2$10,000,000 and $\$^2$0 yields $\$^2$6,666,666,667. The mean squared error is now non-zero, representing the error the model predicts. It can be difficult to interpret the mse values but the metric is still popular. In this case, it may be counterintuitive that the mse is higher than any of the movies' revenue. Generally the mse significantly penalizes predictions that are extremely different than the actual. This is due to the exaggerating impact of the squaring step.

Another popular metric is root mean squared error (RMSE) in which another step is added to MSE. As you might expect the square root is added to the metric as the final step. In the example, the RMSE would be $81,649. This is the square root of $\$^2$6,666,666,667. The benefit of RMSE is that the number scales back towards an expected value closer to the errors themselves.

Still another popular evaluation technique is the mean absolute error. This more straightforward metric deals with the positive and negative error values by taking the absolute value rather than the square root of the square. Then all absolute error values are averaged to arrive at a final evaluation metric. For the fictitious example, the MAE is 66666 because abs(100,000) + abs(−100,000) + 0 or 200,000 is divided by 3.

The cross validated model is built using MSE, but the next section demonstrates how to calculate both RMSE and MAE for additional context.

```
text.cv<-cv.glmnet(dtm,train.y,alpha=1,family=
'gaussian', type.measure='mse', nfolds=5, intercept=T)
```

Once a model is fit you can construct the lambda and "mse" measure plot. Previous sections of the book explain how to interpret the results in Figure 7.6.

```
plot(text.cv)
title("Movie Reviews predict Revenue")
```

The text.cv model contains cross validated information for lambda values and term coefficients. The results of the model need to be evaluated in the next section before being applied to the test set. This helps to ensure that you are not overfitting.

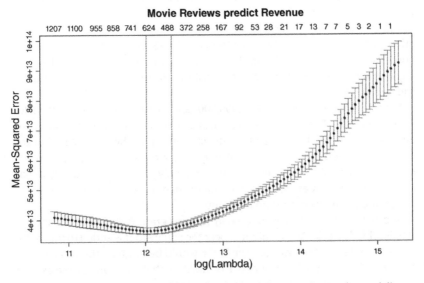

Figure 7.6 Showing the minimum and "1 se" lambda values that minimize the model's mse.

7.3.4   Model Evaluation

The model was fit to minimize the MSE measure and had five-fold cross validation. As a result, calculating and evaluating the model by MSE is redundant. The
code below shows how to calculate the RMSE and MAE for training data. The
explanation for these measures was previously discussed. Remember that at this
step you should adjust model hyper parameters like `alpha` before adding in the
test data and reevaluating the MSE, RMSE and MAE. Adjusting the model parameters after the test set is added could jeopardize the model's true effectiveness.

Using the common `predict` function, pass in the elastic net model then the
training set `dtm`. You need to also specify the lambda value explicitly.

```
text.preds<-predict(text.cv,dtm,s=text.cv$lambda.min)
```

I like to create a simple data frame with the actual training set outcomes and
predictions. This is a (bad) habit and creates redundant data! Nonetheless,
I like to have a standalone data frame in case I want to export it.

```
train.dat<-data.frame(actual=train.y,
preds=text.preds[,1])
```

Using the `train.dat` object, calculate the RMSE by passing in the vector
of true values. The second input is a vector of predicted values. Keep in mind
that the order and number of rows must match. In this case, movies are organized row-wise and the lengths are equal. If you get an error, there is likely an
issue with the way the text was organized and preprocessed. In this example,
the RMSE is 3,179,086. Therefore, on average, the model predictions are ±$3m
from the actual value. Remember that the distribution of actual movie revenues
is non-normal with some outliers.

```
rmse(train.dat$actual,train.dat$preds)
```

Calculating the MAE is also straightforward. The `mae` function accepts the
same data in the same order as `rmse`. In this example, the mean absolute error
is 2,595,979.

```
mae(train.dat$actual,train.dat$preds)
```

To visualize the model effectiveness you can create a box plot of the distributions between actual and predicted values. An easy way to organize the data for
visualization is to convert the `train.dat` data frame into a tidy format. This
transforms the data from a row with prediction and actual revenue columns
into one row per combination. That is to say, one column contains a factor
"actual" or "preds" and the second column contains the value. With this tidy
function, the data loses the association to a specific movie and instead is organized by class of actual or prediction. While the data object is actually redundant, this step demonstrates a quick method for visualization. Table 7.8 shows

Table 7.8 The original `train.dat` data frame.

Row	Actual	Preds
1	27,546	4,139,602
2	3,880,270	3,423,786
3	9,850	4,269,664
...	...	...
1600	12,401,900	15,586,575
1601	23,624,548	17,080,099

Table 7.9 The tidy format of the same table.

Key	Value
Actual	27,546
Actual	3,880,270
Actual	9850
...	...
Preds	4,123,685
Preds	13,080,178
Preds	4,551,006

a portion of the original `train.dat`. Table 7.9 is a tidy version of the same information.

```
train.tidy<-gather(train.dat)
```

Using the tidy version of the predicted and actual movie revenues, you can construct a box plot to compare distributions. The code below refers to the `train.tidy` data frame. To compare distributions make the key vector the x value and y the revenue values. The other parameters apply a visual type and a default set of aesthetics. The box plot visual is captured in Figure 7.7.

```
ggplot(train.tidy, aes(x=key, y=value, fill=key)) +
geom_boxplot()+theme_gdocs()
```

In the box plot you can see that the model has a more narrow prediction distribution. The mean values, represented as the dark horizontal line in the box, are at a similar level. This means that the model is predicting the average movies well, but failing to identify low revenue movies. In the case of outlier movies, the model appears to be under predicting revenue. There are likely ways to improve the model accuracy for non-average revenue. Some examples would be to create a separate model for identifying outlier movie revenue based on new information such as actors, production budgets and directors.

Another visual that can aid in determining the relationship between a model's predicted and actual values is a simple scatter plot. Ideally the scatter plot would show a 45° trend line between the values with less dispersion. Figure 7.8 is the scatter plot created using the `ggplot` code below. To retain the row-wise relationship to a single movie refer to the original `train.dat` data

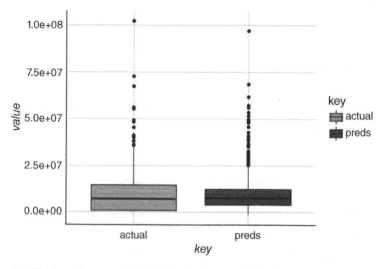

Figure 7.7 The box plot comparing distributions between actual and predicted values.

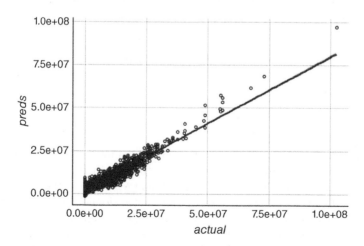

Figure 7.8 The scatter plot between actual and predicted values shows a reasonable relationship.

frame. Instead of passing in the `geom_boxplot` layer the `geom_point` layer is used along with `stat_smooth` for the trend line.

```
ggplot(train.dat, aes(x=actual, y=preds)) +
  geom_point(color='darkred',shape=1) +
  stat_smooth(method=lm) + theme_gdocs()
```

You should apply the model to test data once you are satisfied with the evaluation metrics and have tuned the model parameters to maximize the accuracy. The next section demonstrates how to apply the model to new movie reviews. As you learn more machine learning techniques the text organization and DTM matching is foundational no matter the supervised learning approach that is applied.

7.3.5 Apply the Model to New Movie Reviews

Suppose you are now comfortable with your model and would like to put it into production. As new movie reviews are written you would like to apply the model and get a predicted opening weekend revenue To do so, each new review needs to be cleaned in the same manner as the training text.

```
test.text<-review.clean(test.movies)
```

After the new movie reviews are cleaned you can create an iterator similar to before. The `itoken` function is from `text2vec`, and allows you to work on large corpora as the function iterates through the text rather than doing it all at once.

```
test.it<-itoken(test.text, preprocess_function =
tolower, tokenizer = word_tokenizer)
```

Next you create a new DTM using the iterator and the original training data vectorizer. The `create_dtm` accepts the new text's iterator information, and the original vectorizer ensures that the terms are matched.

```
test.dtm<-create_dtm(test.it,vectorizer)
```

At this point, you can make predictions using the `text.cv` model applied to the `test.dtm` matrix. Remember that in an elastic net you must also specify a lambda parameter.

```
test.preds<-predict(text.cv,test.dtm,
s=text.cv$lambda.min)
```

In this test set example, you know the actual opening weekend revenue. For truly new movie reviews you would have to wait to check the accuracy. However, here you can compare the RMSE and MAE, using the code below. One should expect the evaluation metrics to deteriorate slightly on unseen observations but that the predictions are close and relatively stable. Your RMSE and MAE values will vary slightly, based on the random training data split, but they are likely to be close to an RMSE of $6,479,658 and MAE equal to $4,859,730. In this educational example, there is significant deterioration on

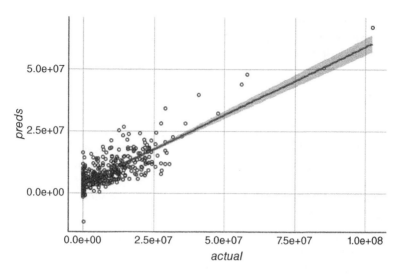

Figure 7.9 The test set actual and predicted values show a wider dispersion than with the training set.

the test set metrics, so you may want to explore ways to improve the data by including using more information.

```
rmse(test.y,test.preds)
mae(test.y,test.preds)
```

The code below visualizes the fact that the model performs less well on test data. In Figure 7.9, the relationship is still observed but the relationship between actual and predicted values has a wider dispersion. As a text miner, if you were still satisfied with the model you could proceed to put it into production knowing that the predictions are slightly under predicting values. If you needed more accuracy, you would feature engineer more information such as release weekend or specific actors to improve the text-only predictions.

```
test.dat<-data.frame(actual=test.y,
preds=test.preds[,1])
ggplot(test.dat, aes(x=actual, y=preds)) +
geom_point(color='darkred',shape=1) +
stat_smooth(method=lm) + theme_gdocs()
```

7.3.6 Movie Revenue Conclusion

This example demonstrates a way to extract and organize text to make continuous predictions. Movie production companies explore data science to identify good bets and avoid costly mistakes. The method shown here is simple yet applies the principles of both text mining and machine learning in a clear manner. If you were to pursue this model further you may want to stack predictions as inputs into a new model while incorporating other known or expected revenue impacting movie attributes.

7.4 Summary

In this chapter you learned how text mining can improve classification methods and also be used as inputs for predictive modeling. Beyond patient readmission, classification algorithms using text can be found in customer service, and fraud identification contexts. Predictive modeling techniques incorporating text are used in finance and marketing in addition to social media.

Specifically, in this chapter you learned:

- how classification is different from continuous prediction
- how to create a DTM using `text2vec` instead of the `tm` package
- how to apply the training set vocabulary to new text
- how to create a classification algorithm for hospital readmissions, based on discharge text and patient information
- how to create a continuous predictive model of social media shares based on article text.

8

The OpenNLP Project

In this chapter, you'll learn

- what is the OpenNLP Project
- the basics of R's OpenNLP package
- an example of syntactic parsing
- what is named entity recognition (NER)
- to load NER libraries
- to perform NER
- to use an API to get latitude and longitude of recognized locations
- to create a bar chart of recognized organizations
- to use a heat map to understand the entity interactions
- to apply polarity scoring to individual entities in a box and whisker plot
- to chart document polarity over time for specific entities.

8.1 What is the OpenNLP project?

The OpenNLP library is a toolkit for supporting natural language processing tasks. It is part of the Apache Software Foundation and is offered for free, much like R. The Apache Software Foundation supports public software creation using defined processes, ensuring consensus among contributors and exceptional quality standards. The OpenNLP project is one of more than 350 projects covering various topics such as big data, databases and email.

The OpenNLP toolkit provides many tools for performing tasks that we have reviewed in this book. For example, tokenization of words can be done in R and also using the OpenNLP Apache library. There are some tasks, however, that OpenNLP performs that many other R text processing libraries do not usually accomplish. The OpenNLP library can be used for part of speech tagging, and named entity recognition. In Chapter 2, you learned about the contrasting approaches between the bag of words and syntactic parsing text mining

Text Mining in Practice with R, First Edition. Ted Kwartler.
© 2017 John Wiley & Sons Ltd. Published 2017 by John Wiley & Sons Ltd.

methods. The OpenNLP project is able to perform many of the syntactic parsing approaches.

However, the openNLP project is written in Java and is therefore not easily accessible to R programmers. Luckily there is an R package that wraps the Java functions so that they can be called using R's syntax in your R session.

8.2   R's OpenNLP Package

Figure 8.1, repeated from Chapter 2 shows a simple sentence being broken down according to syntactic parsing methods. At the time, you learned how a sentence is recognized as a sentence and then broken down into another tag such as a noun or verb phrase and so on. The OpenNLP project does not refer to the word classes as "tags" but as "annotations." In the example below, an annotation model must be used at each level of the process to break down the sentence to its various parts. Annotation models are individually applied to a document for sentences, phrasing, part of speech and named entity recognition.

The OpenNLP package contains only five functions representing the methods of annotating a sentence. The package is not overwhelming in scope but the brief package documentation and lack of wide online usage makes it difficult to figure out for a novice R programmer. The five functions are listed in Table 8.1 along with a short description. The rest of the chapter will apply the functions to a corpus, with explanations so that you can apply the methods and add syntactic parsing to your text mining toolkit.

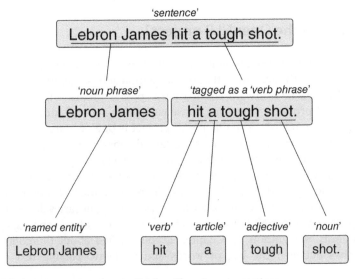

Figure 8.1 A sentence parsed syntactically with various annotations.

Table 8.1 The five functions of the OpenNLP package.

Function name	Description
Maxent_ Chunk_ Annotator	Text chunking identifies the noun or verb groups within a sentence. Chunking classifies an entire chunk, not the structure within a chunk or its relationship to the sentence.
Maxent_ Entity_ Annotator	The name finder can detect named entities and numbers in text. Using the entity annotator requires a pre-trained model specific to a language. Entity recognition needs to have sentence and word annotations applied beforehand.
Maxent_POS_ Tag_Annotator	The part of speech tagger marks word tokens with a class such as noun or adjective. A token can have multiple POS tags, based on probability and relationship to other word tokens. The POS tag codes come from the Penn Treebank Project, and are listed in Table 8.2.
Maxent_Sent_ Token_ Annotator	The OpenNLP sentence detector detects punctuation marks to determine the end of a sentence. This segments each individual sentence in a text so that other more granular annotation models can be applied.
Maxent_Word_ Token_ Annotator	This model segments characters into tokens, often words, numbers or punctuation marks.

The POS tags come from the Penn Treebank Project. The project dates back to the early 1990s and contains syntactic and semantic information on real text examples. The project created tag codes now used in the POS OpenNLP model. The list below will help you decipher the tag codes if you perform POS tagging as part of your project.

It is important that you apply the annotators in the correct order. If you do not, your results may vary from reality. For example consider two sentences below.

His name is George. Washington is where he is from.

If you were to apply an entity extraction annotation model before a sentence annotation model, the model may identify George Washington in the text. However, George is a separate individual from the meaning of Washington. Washington is not used in the context of a last name. Instead sentence annotation is needed first to break up the individual sentences. Applying the sentence annotation model and then the entity recognition annotation will improve results. This is because the entity annotator will identify George as part of the first sentence. It will also identify Washington as part of the second sentence representing some form of entity, possibly a location. The sentence annotation separates the words for the entity annotation so that two entities are split. As you progress through the example script and also explore on your own, remember that the order of annotators is important.

Table 8.2 The Penn Treebank POS tag codes.

POS Tag	Description	POS Tag	Description
CC	Coordinating conjunction	PRP$	Possessive pronoun
CD	Cardinal number	RB	Adverb
DT	Determiner	RBR	Adverb, comparative
EX	Existential there	RBS	Adverb, superlative
FW	Foreign word	RP	Particle
IN	Preposition or subordinating conjunction	SYM	Symbol
JJ	Adjective	TO	To
JJR	Adjective, comparative	UH	Interjection
JJS	Adjective, superlative	VB	Verb, base form
LS	List item marker	VBD	Verb, past tense
MD	Modal	VBG	Verb, gerund or present participle
NN	Noun, singular or mass	VBN	Verb, past participle
NNS	Noun, plural	VBP	Verb, non third person singular present
NNP	Proper noun, singular	VBZ	Verb, third person singular present
NNPS	Proper noun, plural	WDT	Whdeterminer
PDT	Predeterminer	WP	Whpronoun
POS	Possessive ending	WP$	Possessive whpronoun
PRP	Personal pronoun	WRB	Whadverb

Installing the OpenNLP package is straightforward using the code below. However, the entity models require a different download location. For many packages, using an official CRAN mirror can be done using the `install.packages` function without specifying a server location. To download a specific named entity annotator you have to specify a location, because the models are not part of the CRAN repository. An existing model can be found at the Vienna University of Economics and Business. The code below installs the CRAN openNLP library and the Vienna University annotation model.

```
install.packages('openNLP')
install.packages("openNLPmodels.en",repos = "http://
datacube.wu.ac.at/",type = "source")
```

What is Maxent (maximum entropy)?

The annotation models use a method called maximum entropy to identify the sentences, tokens, parts of speech and named entities. Maximum entropy is a machine learning technique for classification. The dependent class is the annotation tag such as noun. The input features are patterns about the text that have been observed. However, maximum entropy is an information-based technique, not a probability-based technique. Probability counts the individual occurrences of features, while information is based on the average pattern occurrence for a feature. This is well suited to natural language because of expression diversity. In natural language, there are many specific probability distributions so a tagging model has to be more generally applied to be useful. Maximum entropy allows you to be less specific but also less error prone than providing an explicit probability distribution for each noun, verb and so on.

Entropy is a measure of uncertainty in a distribution. Entropy measures an event probability and the event's "surprise," given the prior observation's average occurrence. In common terms, if you thought winning the lottery was unlikely then your surprise would be very high if you had the winning ticket. In contrast, flipping a coin and getting "tails" is less surprising because you would expect a more likely chance of that outcome. The equation for entropy is as follows.

$$x \quad = \quad Event$$
$$p_x \quad = \quad Probability$$
$$Surprise \quad = \quad \log(1/p_x)$$

$$Entropy, Expectation\ over\ Surprise = -\sum_{x \in X} p_x log_2 p_x$$

For annotation models, entropy (uncertainty) is maximized, while still resembling the reference word distribution. Rather than having complete uncertainty and a useless model, the model uses average probability constraints to lower the entropy. The maximum entropy approach balances complete uncertainty by applying constraints based on observed information.

For example, to define a sentence you need punctuation marks, capital letters and whitespace constraints. This model allows for uncertainty in the specific punctuation, letter or amount of whitespace to identify the sentence. Thus, the uncertainty is maximized given what you know about sentences but is not explicitly defined using a period, capital "T" and two spaces afterwards because sentences can end in exclamation points, start with another letter or have a single blank space after.

Another common explanation leverages your own reading ability. In the text below, you identify a word solely on the first and last letters. Your reading

> "annotation model" uses the minimum characteristics of a word to identify the word. In doing so, your mind's algorithm allows for maximum entropy for the other characters in the words.
>
> "Raednig wrdos is esaeir tahn you tuohgt. Tnurs out, olny the fisrt and lsat lteters mttater. Yuor mnid is mxaizimnig the ucnectriatny to cpermohned the wrods."

Once it is installed you are ready to begin a case study examining hundreds of emails using syntactic parsing.

8.3   Named Entities in Hillary Clinton's Email

Figure 8.1, repeated from Chapter 2, shows a simple sentence being broken down according to syntactic parsing. Now you are going to apply this approach to hundreds of emails. During the US presidential election, the Democratic Party nominated Hillary Clinton. While she was the Secretary of State under Barack Obama, she used a personal email server. During her candidacy for president, the opposition party launched an investigation into whether classified information was not properly secured, which constitutes a crime. As part of the government investigation, thousands of emails were released to the public. Regardless of political affiliation or feelings on this divisive issue, the emails pose an interesting corpus to demonstrate syntactic parsing and named entity recognition.

Following the text mining workflow, suppose you are a journalist looking to understand the emails quickly. If this were the case, you would define the problem statement in an exploratory manner.

1) **Define the problem and specific goals.** Explore Hillary Clinton's emails to automatically identify people, places and organizations. Then perform polarity analysis on individual entities to understand how Hillary Clinton's language changes among different entities.
2) **Identify the text that needs to be collected.** Some 25,000 individual emails were released to the public as official documents. These emails were released in chunks throughout the inquiry. In this case study, you will examine 551 emails released in Feb 2016. This is a small subset of the larger email collection but is enough for educational purposes. If you were performing the analysis on all emails it would require more RAM and likely take longer to analyze.
3) **Organize the text.** This analysis requires systematically reading individual files to construct a larger single corpus comprising 551 documents. As part of the organization step, some rudimentary cleaning functions are applied. If desired, you can apply more robust preprocessing methods learned in earlier chapters.

Getting the Data

Please navigate to www.tedkwartler.com and follow the download link. For this analysis, please download "C8_final_txts.zip". The zipped file contains 551 individual emails from the Hillary Clinton investigation as plain text files. If desired, the other emails in the investigation can also be downloaded. The code in this chapter assumes that you have unzipped the file and placed all the text files in a single folder.

---

4) **Extract features.** The openNLP annotation models will identify individual sentences, words and ultimately persons, locations and organizations.

5) **Analyze.** In this case study, you will construct multiple visuals. First, you create a simple bar plot to identify the frequent organizations. Next is a world map of recognized locations. Lastly, polarity scores add another layer to the entity exploration. In your own named entity project you can change the entities to construct different plots and analyses such as word clouds or treemaps.

6) **Reach an insight or recommendation.** The goal is largely exploratory in this example. However, this method can be used by marketers to identify named entities in social media text. In this case you are automatically identifying and quantifying various named elements and then focusing on how Hillary Clinton's language changes in relation to specific entities.

8.3.1 R Session Set-Up

This is a complex analysis requiring multiple packages. After downloading the Vienna University annotation model you can call it using the code below. This section sets your working directory containing the folder with 551 individual text files. The code then sets global session options similar to previous examples. Next you load ggmap and ggthemes. The ggmap package provides access to the Google Maps API for constructing ggplot2 visuals. The ggthemes library provides quick aesthetic palettes within ggplot2 illustrations. When loading these packages ggplot2 will automatically load, so you do not have to load it explicitly. After these packages, load openNLP and the Vienna University annotation model, openNLPmodels.en. Although specifying the exact folder location of the annotation model may not be needed, the code below uses an explicit file path. It is important to load the ggplot2 related packages before the openNLP library. Both share a function called annotate. If the order is reversed, the openNLP annotate function is overwritten and your code will not function. Next pbapply, the progress bar version of the apply functions, is loaded. This is merely a convenience to understand the time it will take for some of the longer processing steps. The stringr package provides string manipulation functions. Rvest is primarily used for web scraping but is used here to select specific elements from a list. The doBy package provides group-wise statistics functions that can be helpful when working with many data elements. Once again the tm package is loaded. Instead of using the package for a bag of words style analysis, the package is used for preprocessing functions. Lastly, cshapes provides shape files for maps. In this case, the shape files are used for creating a world map.

```
options(stringsAsFactors = FALSE)
Sys.setlocale('LC_ALL','C')
library(gridExtra)
library(ggmap)
library(ggthemes)
library(NLP)
library(openNLP)
library("openNLPmodels.en", lib.loc="~/R/
win-library/3.2")
library(pbapply)
library(stringr)
library(rvest)
library(doBy)
library(tm)
library(cshapes)
```

At this point, your R session is set up with appropriate options and libraries. Next you need to scan your working directory for the multiple files representing Hillary Clinton's emails. To do so, use list.files along with a pattern within the files. Pass in a wildcard (*) with ".txt" to get a string vector of any files names in the working directory that have the .txt file extension. You can change the pattern to identify other file types or names. Once created, the temp object of 551 file names is passed into an "if" statement. The "if" statement uses a temporary variable "i" along with two functions, assign and readLines. The "if" statement works by iterating from 1 to the length of temp (551), assigning an object name corresponding to the temp file name. Specifically, the assigned object name is created using readLines which is the base R function for reading in text lines. For example, the first file name, temp[1], is "C05758905.txt" This is passed to the "if" statement into the other functions as assign("C05758905.txt"), readLines("C05758905.txt"). The "if" statement then moves to temp[2], C05758988.txt, passing the string into both functions and so on for all 551 files. At this point, you have 551 individual objects in your R session. This can be cumbersome to deal with, so the next line uses pblapply along with get to organize the files into a single list with 551 elements.

There are more succinct ways of organizing multiple files into a single object, but separating the if and pblapply functions makes learning the procedures straightforward.

```
temp <- list.files(pattern='*.txt')
for (i in 1:length(temp)) assign(temp[i],
readLines(temp[i]))
all.emails<-pblapply(temp, get)
```

8.3.2 Minor Text Cleaning

The emails are well organized with few to no misspellings, emoticons or special characters. Thus, the cleaning function can be fairly short. As you apply the txt.clean function in the next code section to other OpenNLP text mining projects you should adjust the preprocessing functions. To begin, the txt.clean function accepts a character string and drops the first line. The first line contains information that is not needed for a named entity analysis. The first line has been added by investigators and is not part of the actual email. An example first line is below.

UNCLASSIFIED U.S. Department of State Case No. F-2014-20439 Doc No. C05758905 Date: 02/13/2016

```
txt.clean<-function(x) {
 x<-x[-1]
 x<-paste(x,collapse= " ")
 x<-str_replace_all(x,
 "[a-zA-Z0-9_.+-]+@[a-zAZ0-9-]+\\.[a-zA-Z0-9-.]+", "")
  x<-str_replace_all(x,
  "Doc No.","")
 x<-str_replace_all(x,
  "UNCLASSIFIED U.S. Department of State Case No.","")
 x<-removeNumbers(x)
 x<-as.String(x)
 return(x)
}
```

After dropping the first line, each email is collapsed into a single text line representing the entire email. When using readLines each line break within an email denotes a new row of the text vector. Collapsing the lines into a single line makes the analysis coincide with 551 emails instead of thousands of individual lines of the email corpus. Next a regular expression to remove email addresses is applied. Using stringr's str_replace_all function with [a-zA-Z0-9_.+-]+@[a-zA-Z0-9-]+\\.[a-zA-Z0-9-.]+ will match emails for pattern replacement. Emails follow a predictable pattern of letters and/or numbers then the "@", a domain made of letters or numbers and finally a .com, .net or some other domain pattern. The regular expression looks for letters or numbers before and after the "@" along with a period followed by any other letters or numbers. Once recognized, the pattern is replaced with an empty character because there is no space in between the quotation marks following the regular expression. Additional string replacements are also performed for other terms that have been appended by the investigative team.

`RemoveNumbers` from `tm` is applied next. Sometimes, the named entity models identify a year as an organization. This function simply removes the numbers, but this may not always be appropriate in other use cases. Lastly, the function changes the email text from character to an object class "string" used by openNLP. The string class is used because it has the ability for subscripting by start and end character position similar to indexes in data frames.

Using `pblapply` along with the list of emails and the text cleaning function will preprocess and change the class of the emails. The returned list is still called `all.emails`. At this point, the text is no longer in a raw state.

```
all.emails<-pblapply(all.emails,txt.clean)
```

To understand the string object class, the code below refers to the third cleaned email from characters 2 to 8. The [2,8] is subscripting to the characters of the string. In this case, it is the word "Subject."

```
all.emails[[3]][18,24]
```

If necessary, you can name the list elements representing individual emails. Naming the list elements makes organization easier in a complex analysis. Pass the `all.emails` list into the `names` function along with the vector of names, "temp." Once it is named, you can refer to list elements by file name and the character start and stop subscripts such as "all.emails$C05759073.txt[2,8]."

```
names(all.emails)<-temp
```

You should remove the 551 text file objects from your workspace to free up space. The individual files are redundant to the clean list elements within `all.emails`. Do not remove the individual objects if you expect to later use the unprocessed versions for other text mining methods. To remove multiple items in your workspace at once, employ `rm`, standing for remove, and passing in a list of objects to remove. In this case, you supply the `temp` vector of file names.

```
rm(list=temp)
```

8.3.3 Using OpenNLP on a single email

To understand named entity recognition you will apply the openNLP functions on a single email before the entire corpus. To begin, you need to call the annotation models. This is done using the function `Maxent_Entity_Annotator`

Table 8.3 The named entity models that can be used in `openNLPmodels.en`.

Kind	Description	Example entities
date	A named entity model to extract dates from text	Thursday April 29, 2011 tomorrow
location	A named entity model to extract possible locations including proper nouns	United States Syria The White House
money	A named entity model to identify currency amounts	$96.5 million $6.18
organization	A model to extract organizations by name or acronym	National Security Council
percentage	An algorithm to identify percent values including numbers, words or using the % symbol.	75 per cent 83%
person	A method to identify proper nouns for people, primarily first and last name	Tony Blair Bill

along with the type of model. Table 8.3 lists the English annotation models within the `openNLPmodels.en` library.

Each entity model such as "person" or "location" incorporates multiple features of the individual token. For example, when reviewing the token "Syria," 13 distinct features are used to identify it as a location. Each kind of entity has a different weight assigned to each of the 13 features. The weights concerning digits are stronger for currency compared to the model for locations. As a result, Syria is not recognized as a currency. Instead, the feature weights for capital letters are made higher in the location model, and Syria is more likely to be a location. Each of the six models is an empirically weighted combination of weights with a binary classification outcome corresponding to the specific entity. The feature list for named entity models includes:

1) token is lowercase
2) token is two digits
3) token is four digits
4) token contains a number and a letter
5) token contains a number and a hyphen
6) token contains a number and a backslash
7) token contains a number and a comma
8) token contains a number and a period
9) token contains a number
10) token is all caps, but is a single letter

11) token is all caps but is more than one letter
12) first letter of the token is a capital letter
13) token was tagged as "other" e.g. (a sentence).

It is time to apply the `maxent` annotations using the downloaded models. The following code creates annotations for persons, locations and organizations. Also, individual models for sentences, words and parts of speech are also set up. These load the pre-existing feature weights to be called by your R session, but they do not yet apply them to any text.

```
persons <- Maxent_Entity_Annotator(kind = 'person')
locations <- Maxent_Entity_Annotator(kind = 'location')
organizations <- Maxent_Entity_Annotator(kind =
'organization')
sent.token.annotator <-
 Maxent_Sent_Token_Annotator(language = "en")
word.token.annotator <-
 Maxent_Word_Token_Annotator(language = "en")
pos.tag.annotator <-
 Maxent_POS_Tag_Annotator(language = "en")
```

You will apply these annotation models to a single email. The code below applies the models using the `annotate` function. Pass in the third email using `all.emails[[3]]`. Then provide a list of the annotation models to be applied in sequence. The sentence annotation model is first followed by the word model and part of speech tagging model.

```
annotations <- annotate(all.emails[[3]],
list(sent.token.annotator, word.token.annotator,
pos.tag.annotator, persons, locations, organizations))
```

The annotations object is a list that is best changed to a data frame. The code below makes a data frame from each list element, and selects the second through fifth columns. Then a new data frame vector, `features`, is added to `ann.df`. The features vector contains the type of identified entity.

```
ann.df<-as.data.frame(annotations)[,2:5]
ann.df$features<-unlist(as.character(ann.df$features))
```

The resulting data frame can be examined using indexing, with accompanying results printed below. The data frame contains the token type, start and end character positions and the specific feature or entity class.

```
ann.df[244:250,]
      type   start  end         features
244 entity  662    670         list(kind = "person")
245 entity  857    867         list(kind = "person")
```

```
246 entity   897      921          list(kind = "person")
247 entity    70       74          list(kind =
"organization")
248 entity   258      262          list(kind =
"organization")
249 entity   516      520          list(kind =
"organization")
250 entity   709      713          list(kind =
"organization")
```

At this point, the data frame contains the starting and ending character for each entity, but it is more useful to append a vector with the actual token. For a single document, you can use a "for" loop to migrate row by row in the data frame. The anno.chars object is created from the loop of one to the number of rows in the data frame. In this sequence from 1 to 251, the loop uses substring on the third email and passes in a corresponding start and end integer from the ann.df data frame. For example, on the ninth entity row in the data frame the code would be substr(all.emails[[3]], ann.df[9,2], ann.df[9,3])). In the loop, the changing row number is represented by i as the sequence continues. This saves writing 251 individual lines of code. The final object is then added to the original data frame as a vector called ann.df$words.

```
anno.chars<-NULL
for (i in 1:nrow(ann.df)) anno.chars[i]<-
((substr(all.emails[[3]],ann.df[i,2],ann.df[i,3])))
ann.df$words<-anno.chars
```

This final data frame contain not only does persons, locations and organizations, but also each detected sentence, word and part of speech. To complete the named entity analysis you may need to subset the data frame for specific features. The code below used the logical grepl as a subset parameter for each entity class.

```
subset(ann.df$words,grepl("*person",
ann.df$features)==T)
subset(ann.df$words,grepl("*location",
ann.df$features)==T)
subset(ann.df$words,grepl("*organization",
ann.df$features)==T)
```

The third email in the collection does not contain any locations, but other entity types are identified. For example, the subset results for "person" and

Table 8.4 Named people found in the third email.

Person
Jack
Jim
Mills
Jacob J
Steinberg
Alexander
Matt Spence
Alex Alex Pascal National

Table 8.5 Named organizations found in the third email.

Organization
Group
Group
Group
National Security Council

"organization" are contained in Tables 8.4 and 8.5 respectively. You will notice some additional data clean up must be performed on the results. The raw results are non-unique and in the case of "Alex Alex Pascal National" the algorithm concatenated two signatures in a row.

As an investigative journalist in this case study, you now would want to identify the specific part of an email containing an interesting entity. In this example, `National Security Council` pops out, so you need to identify the line in the third email containing this organization. Remember that you are only working on a single email at this point. Previously you removed all 551 individual emails so you must reload the third email referencing `temp[[3]]`. Reviewing the existing object `all.emails[[3]]` may not be useful because you already applied some text cleaning processes to it. The `grep` command will return the line number or numbers containing "National Security Council." The last line indexes the third email to only the line containing the pattern and will print it to the console.

```
third.email<-readLines(temp[[3]])
entity.pos<-grep("National Security Council",
third.email)
third.email[entity.pos]
```

The results are shown in the sentence below. It is a signature line from the email with Alex Pascal's contact information.

```
"We apologize for the late notice. Please feel free to
contact me or Matt Spence with any questions. Thanks,
Alex Alex Pascal National Security Council apascal
(202) 456-9491 (main)"
```

8.3.4 Using OpenNLP on Multiple Documents

Now that you understand the principles of applying openNLP models to an individual email, it is time to perform named entity recognition on a larger corpus. The principles are the same, but the code changes slightly. The code in this entire section supports identifying named entities within 551 Hillary Clinton emails to make interesting visualizations.

To start, you need to set up a custom function that can be applied to the entire list of emails. The `annotate.entities` function accepts an individual document and a "pipeline." The pipeline is a list of the models you wish to employ on the document, but the list itself is defined later. Within the function, annotations are made on the doc using the pipeline, and then a new function `AnnotatedPlainTextDocument` is called. The new function takes the annotations and reapplies the identified tags to the original document. The function will return the plain text document with annotations.

```
annotate.entities <- function(doc,
                     annotation.pipeline) {
 annotations <- annotate(doc, annotation.pipeline)
 AnnotatedPlainTextDocument(doc, annotations)
}
```

To define an annotation sequence you create a list with the specific openNLP models. In this example, the sentence, word and part of speech annotation models are the first elements of the list. Next the named entity models person, location and organization are added. You can change the model "kind" or add new ones separated by commas from Table 8.3.

```
ner.pipeline <- list(
  Maxent_Sent_Token_Annotator(),
  Maxent_Word_Token_Annotator(),
```

```
Maxent_POS_Tag_Annotator(),
Maxent_Entity_Annotator(kind = "person"),
Maxent_Entity_Annotator(kind = "location"),
Maxent_Entity_Annotator(kind = "organization")
)
```

Now that you have an annotation function that individually calls the models, you need to apply them to the entire email list. Since `all.emails` is a list of 551 cleaned emails, you can use either `lapply` or `pblapply` to perform your annotations. The progress bar version, `pblapply`, is helpful because named entity modeling can be time-consuming depending on the size of the corpus. The `pblapply` function accepts the email list, `all.emails`, then the custom function that returns annotated plain text documents called `annotate.enti-ties`. Since the `annotate.entities` function also needs the named entity models list you also pass in the `ner.pipeline` list separated by a comma.

```
all.ner<-pblapply(all.emails,annotate.entities,
ner.pipeline)
```

After completing the 551 separate annotations you can extract the useful information and construct a data frame with entity information. The `pluck` function can extract list elements by name or index. The code selects each annotation list from the 551 elements. Then each of the annotations is changed to a data frame. At this point `all.ner` is a list containing all named entities made of 551 elements. Each element of the list is a single data frame corresponding to the email order with a type vector, start and end character position and the specific feature identified. This mimics the single email data frame shown previously.

```
all.ner<-pluck(all.ner,"annotations")
all.ner<-pblapply(all.ner, as.data.frame)
```

Indexing `all.ner` allows you to review the results. The code reviews the third email and the same rows shown in the single email example. The third email's data frame of annotations can be accessed using double brackets and 3. Annotation rows 244 to 250 are selected along with all columns. The row indexing is performed within single brackets. This exactly matches the single email with three person and four organization entities.

```
all.ner[[3]][244:250,]
```

The previous loop function will not work on `all.ner` and `all.emails`. The next code constructs a single data frame with the same information as the previous loop, type, start, end, features and the actual token, but also adds a file name vector. With so many emails the "file" vector identifies where the entity was found.

Instead of a loop, the code now uses the `Map` function, which applies a function to corresponding elements of any vectors or lists it is supplied. Using `Map`, the custom function below accepts text, features and an id object. The custom function uses cbind to create a data frame. The data frame vectors are defined inside the cbind function. The vectors include the original data frame, the substring function to extract the token and the file name. The last part of the Map function contains the objects used within the custom function that are being mapped. For example, all.emails[[1]] will be the "tex" input, all.ner[[1]] represents the "fea" input and temp[[1]] is the "id." The first row of the data frame is mapped to be all.ner[[1]], the characters from substring between all.ner[[1]]$start and all.ner[[1]]$end and temp[[1]]. The temp[[1]] string is automatically repeated for the length of the all.ner[[1]] data frame. This will be mapped over all 551 individual data frames. The result is still a list of 551 emails but with new vectors for the token and corresponding file.

```
all.ner<-Map(function(tex,fea,id) cbind(fea,
    entity=substring(tex, fea$start,fea$end),file=id),
    all.emails,all.ner,temp)
```

Although the Map function is complicated, the end result should be familiar. Calling the third email and the same rows as before shows the additional information that was appended to the data frame. This includes the entity vector with the word token and the corresponding file.

```
all.ner[[3]][244:250,]
      type    start   end    features    entity      file
244   entity   662    670               person      Alexander
C05759013.txt
245   entity   857    867               person      Matt
Spence   C05759013.txt
246   entity   897    921               person      Alex Alex
Pascal National   C05759013.txt
247   entity    70     74    organization            Group
C05759013.txt
248   entity   258    262    organization            Group
C05759013.txt
249   entity   516    520    organization            Group
C05759013.txt
250   entity   709    713    organization            Group
C05759013.txt
```

Using do.call with rbind and all.ner will create a unified entity data frame from the 551 individual data frame list elements. The features vector is actually a nested list class, so you must add code to flatten the vector. The extra step to change all.ner$features into a simple character vector makes data manipulation easier.

```
all.ner<-do.call(rbind,all.ner)
all.ner$features<-unlist(as.character(
  all.ner$features))
```

The `all.ner` object is a single large data frame. Each row represents an annotation for the entire corpus. Each individual sentence, word and part of speech was identified in addition to people, locations and organizations. In this case study, the 551 emails contain 242,585 annotations. As a result, the `all.ner` data frame has 242,585 rows and seven columns.

Indexing to only the data frame rows for particular entity classes will make the data more manageable. The grep function will return row integers matching a pattern and can be nested inside the `all.ner` indexing brackets. This code creates three small data frames with seven columns. The three smaller objects correspond to the person, location and organization model entities. Remember to change the pattern if you are using a different entity model such as "money."

```
all.per<-all.ner[grep("person", all.ner$features),]
all.loc<-all.ner[grep("location", all.ner$features),]
all.org<-all.ner[grep("organization",
                all.ner$features),]
```

8.3.5 Revisiting the Text Mining Workflow

Before moving on to the next section, let's quickly revisit the text mining workflow. To this point, you have completed steps 1 through 4. Step 1, the problem statement and goal, has not changed and is restated here. To satisfy step 2, you downloaded the raw text identified for the case study. Your previous code executed step 3 to organize the text into a list and performed some basic cleaning. Finally, in step 4 you ultimately extracted three objects that can be used for analysis.

1) **Define the problem and specific goals.** Explore Hillary Clinton's emails to automatically identify people, places and organizations. Then perform sentiment polarity on individual entities to understand how Hillary Clinton's language changes among different entities.
2) **Identify the text that needs to be collected.** You downloaded 551 Hillary Clinton emails released in Feb 2016.
3) **Organize the text.** You have systematically read and organized the emails so that openNLP models can identify entities.
4) **Extract features.** All named entities for people, locations and organizations have been extracted from the text using the openNLP models. The entity features are organized into individual data frames for people, locations and organizations.

8.4   Analyzing the Named Entities

One of the first things you may want to do is identify the unique entities in each of the three data frames. The `unique` function returns a vector, data frame, or list like the object it was passed. The difference is that the duplicate values, rows or elements are removed. Here, you create a vector of 370 unique locations from the `all.loc$entity` character vector.

```
uni.loc<-unique(all.loc$entity)
```

Another method for obtaining similar information is with the function `firstobs` from the `doBy` package. The `firstobs` function identifies the first instance of a unique value and returns its position. A similar function is `lastobs`, which will identify the position of the last unique observation. Used in the code below, the first unique observation indexes location entity vector. Both methods result in 370 unique observations but `firstobs` identifies the first mention of an entity. This can be helpful if your data is chronological and you want to know at a specific point that an entity is used. To explore different entities, change `all.loc` to another of the three entity data frames.

```
uni.loc<-all.loc[firstobs(~entity, data=all.loc),]
```

Another basic data exploration technique is to perform a frequency analysis. A simple bar plot of the top unique terms can be insightful. A fast method to summarize the entity text is to create a table of the entity information. Table needs to be applied to a factor, not character string, so you must first change the vector to a categorical variable. The resulting object class also needs to be changed from table to matrix. All three functions are nested and applied to the organization entities below. Once again, change the data frame entity to explore persons or locations.

```
orgs<-as.matrix(table(as.factor(all.org$entity)))
```

Once created, the orgs matrix has 871 different organizations with a corresponding frequency integer. You can explore the matrix by index position with the results in Table 8.6.

```
orgs[216:220,1]
```

A bar plot will be incomprehensible with 871 different values. One method to reduce the information is to reorder the entire matrix using `order` and then select the most frequent terms. First, the code sorts the matrix in a decreasing manner. After reordering, the plot's side margins are explicitly defined so that the text is not cut off. Adjust the integers accordingly in your own plot as text lengths vary. Next the base function `barplot` constructs the visual by accepting the matrix indexing the top 20 terms. The `las` parameter adjusts the bar labels to be vertical. Finally, the resulting bar plot is illustrated in Figure 8.2.

Table 8.6 Example organizations that were identified and their corresponding frequency.

Organization	Frequency
Department of Justice	2
Department of State	2
Department of State All	1
Department of State Washington	2
Department of Treasury	1

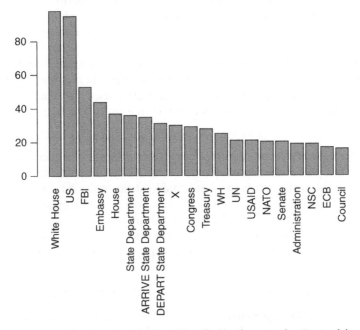

Figure 8.2 The most frequent organizations identified by the named entity model.

```
orgs<-orgs[order(orgs[,1], decreasing=T),]
side.margins <- par(mar = c(11,2,1,1) + 0.3)
barplot(orgs[1:20], las = 2)
```

8.4.1 Worldwide Map of Hillary Clinton's Location Mentions

After exploring the unique, first and last observed terms and frequency for the three entity classes, you may want to construct a world map and plot unique locations. Once again using unique, creates a location character vector with the code below.

```
uni.loc<-unique(all.loc$entity)
```

To plot the locations, this code section uses the Google Geocoding API. Geocoding is the process of identifying the geographic location for a specific place. The ggmap library provides automatic bindings for Google's Geocoding API service. You can use the service to obtain 2500 latitude and longitude pairs a day for free.

The geocode function accepts a character vector of places. Assuming that you have an Internet connection, you pass in the unique locations to obtain all the latitude and longitude pairs. If the API service fails to recognize a location, an error is returned for that record, but the function will continue to check other places. Each response is formatted as JavaScript object notation (JSON) packets, but the geocode function conveniently handles parsing the information. The geocoding will be complete after a few moments and multiple updates to your R console.

```
uni.loc.geo<-geocode(uni.loc)
```

You will want to column bind the coordinate pairs to the original data. That way you can save it for later, without having to redo the time-consuming geocoding. Using cbind, the uni.loc object contains the original location information along with corresponding latitude and longitudes. If a location was not recognized the values will be "NA."

```
uni.loc<-cbind(uni.loc,uni.loc.geo)
```

There are multiple methods to construct a worldwide map. One method is to use the get_map function from ggmap. This queries the Google Map API service and returns a static image to be used with ggplot2 functions. A different method is to use shape files defining the country boundaries on a map. Loading the cshapes package loads the functions and underlying data to construct a simple worldwide map. The shape file method is used to construct this visual.

The innermost function in the code below is cshp, which is a that function extracts the worldwide country data as of a particular date. The shape file data is passed to the fortify function from ggplot2. Fortify will construct a data frame containing latitude and longitude points with grouping identification numbers. The polygon data will become the base layer for the worldwide map. Note that the column names include long, lat and group, which will be used later.

```
world.data <- fortify(cshp(date=
                      as.Date("2015-06-30")))
```

To construct the base layer, call ggplot below. Pass in the world.data data frame, and define the aesthetics with x,y and grouping. The aesthetics will correspond to longitude, latitude and country group, according to the data frame column names "long," "lat" and "group." The next line adds geom_polygon to apply a fill and border color with weight. The base map, Figure 8.3, is plotted with

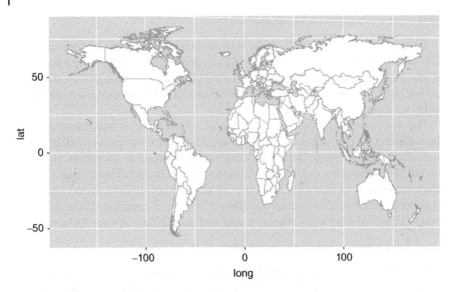

Figure 8.3 The base map with ggplot's basic color and axes.

ggplot2's basic aesthetics. The base map does not contain the location entities yet.

```
base.map <- ggplot(world.data,
aes(long,lat,group=group)) +
  geom_polygon(fill="white", color="grey80",
size=0.25)
```

The email.locs layer adds geometric points using the uni.loc data. Using geom_point treats the next layer like a simple scatter plot on top of a world map. As before, the aesthetics are defined with latitude and longitude, but the data frame names have changed to "lon" and "lat." In this case, there is no grouping because the points are unique. Next define size, transparency and color for each point. To quickly change the ggplot default background and axes, use a predefined theme from ggtheme. The code uses "theme_few" because it is simple. Using predefined ggthemes is a quick way to change multiple aspects of any ggplot2 visual. Lastly, add a title using ggtitle and call the plot. The final map is shown in Figure 8.4.

```
email.locs<-base.map+geom_point(data = uni.loc,
             aes(x = lon, y=lat, group=NA), size=2.0,
             alpha=.5, colour="red")+
  theme_few()+ggtitle("NER Locations")
email.locs
```

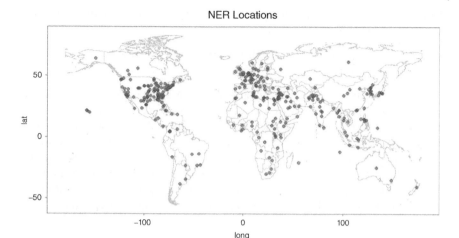

Figure 8.4 The worldwide map with locations from 551 Hillary Clinton emails.

Using semi-transparent dots by adjusting the alpha parameter allows your audience to gauge the intensity of the location mentions. Nearby areas will overlap and become darker while infrequent locations will have more transparent points.

Some locations are assigned a generic point representing a larger region by the API service. For example, the location "Russia" is assigned coordinates in central Russia, although the Secretary of State would likely be talking about the Russian government based in Moscow. If the misplaced points materially affect your analysis, you should manually review the uni.loc character vector prior to geocoding and plotting.

Also, ggplot2 will automatically remove any locations that have NA for longitude and latitude values. NA values are worth investigating because the locations may be misspelled and therefore unrecognized by Google's API service. In this example, 17 locations were removed before plotting Figure 8.4.

8.4.2 Mapping Only European Locations

Another way to construct a map is to use Google's static map API. The ggmap library provides an easy way to obtain a map. Once you have the map, it is used as the base layer of a normal ggplot2 illustration. Once it is the bottom layer, you can plot points on top with the same code that was used in the world map.

One method to call the Google service is with get_googlemap. This code uses Luxembourg as the map center because the nation is centrally located to the continent. If desired, change the location string to recenter the map. The next parameter ranges from 3 to 21. The default zoom is 10 which is the city level for a map. The last parameter contains the map type that is returned from

Google. The options are roadmap, satellite or hybrid. The static map information is stored in the object `eu.map`.

```
eu.map <-get_googlemap('Luxembourg', zoom=4,
maptype='roadmap')
```

The `eu.map` information is then plotted as the foundation for your visual. Use the function `ggmap` along with the returned API object. Adding "`extent = device`" to the image extends the map to the edge of the graphics window and removes latitude and longitude axes. Finally, add a layer of individual locations using `geom_point` with the same aesthetics as before. Once it is created, ggplot will produce a warning for any points that do not fit on the map or have NA values. This is helpful since ggplot automatically omits the points that do not fit within the map. Figure 8.5 is Google's European map with the email locations plotted.

Figure 8.5 The Google map of email locations.

```
ggmap(eu.map,extent = "device")+
  geom_point(data = uni.loc,
             aes(x = lon, y=lat, group=NA),
             size=2.0, alpha=.5, colour="red")
```

Another function get_map allows you to change the map appearance further. In the code below, the get_map function fetches Google's Luxembourg information. The additional parameters, source='stamen' and maptype='toner', download black and white map tiles from www.stamen. com. These tiles are assembled in order to produce the European continent. After assembly, the same code adds the layer containing email locations. Figure 8.6 is the black and white version of the European map.

Figure 8.6 Black and white map of email locations.

```
bw.map <- get_map('Luxembourg',
zoom=4,source='stamen', maptype='toner', crop=F)
ggmap(bw.map,extent = "device")+
                  geom_point(data = uni.loc,
                        aes(x = lon, y=lat, group=NA),
                        size=2.0, alpha=.5,
colour="red")
```

8.4.3 Entities and Polarity: How Does Hillary Clinton Feel About an Entity?

A potentially insightful visual could incorporate sentiment polarity distributions for specific named entities. Comparing the distributions may indicate the overall positive or negative attitude of an author in relation to other entities. This section identifies some specific named entities, reloads the original emails, applies qdap's polarity function and ultimately creates a "box and whisker" plot.

A box and whisker plot is a compact visualization to understand individual distributions. The values of a distribution are illustrated vertically. Between the first and third quartile, a box is created, giving the illustration its name. A thick horizontal line is drawn within the box. The line represents the median value of the distribution. Another line extends from the top and bottom of the box. These "whiskers" represent the maximum and minimum values excluding outlier values. Lastly, individual dots are placed beyond the whiskers representing any outliers. Outliers are usually defined as being 1.5 times the upper and lower quartiles. Box and whisker plots are a visual representation of the shape of a data set. Figure 8.7 shows a normal distribution and a box and whisker plot side by side. The box and whisker plot has been rotated so you can understand the relationship between the two.

Figure 8.7 is a single normal distribution. As a result, it is not very informative as a box and whisker plot. More exciting conclusions can be drawn when there is more than one box and whisker plot representing a category and the distributions are skewed.

As you become more familiar with the data you can identify interesting named entities. Sometimes this is due to a large number of mentions or because you did not expect the entity in the corpus. For this example, you will create a bar plot for emails containing "Russia," "Senate" and "White House." As you have seen before, using grep will identify the row positions of the specific patterns. The integers from the grep commands are then used to index the seventh column of the various person, location and organization entity data frames. In your own work, remember that the data frame used in the grep step must coincide with the index code. Otherwise the code will select the wrong emails. For

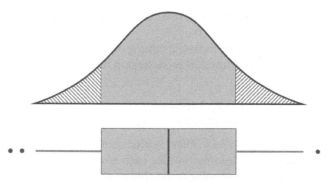

Figure 8.7 A normal distribution alongside a box and whisker plot with three outlier values.

example, using grep on all.org$entity, must be passed into the indexing code as all.org["grepped integers",7].

```
senate<-grep("Senate", all.org$entity,ignore.case = T)
white.house<-grep("White House",
all.org$entity,ignore.case = T)
russia<-grep("Russia", all.loc$entity,ignore.case = T)
se.files<-all.org[senate,7]
wh.files<-all.org[white.house,7]
ru.files<-all.loc[russia,7]
```

The three entity file objects are concatenated into a single text vector. The vector now contains all file names for all three entities.

```
three.ent.files<-c(se.files,wh.files,ru.files)
```

At the beginning of the case study a loop was used on the temp object to load 551 individual emails. The same code can be applied to the smaller file name vector. The files are organized into a list using the get function with pblapply. Finally, the rm function removes the individual file objects.

```
for (i in 1:length(three.ent.files)) assign(
  three.ent.files[i], readLines(three.ent.files[i]))
three.ent.emails<-pblapply(three.ent.files, get)
rm(list=three.ent.files)
```

Although it is inadvisable, the standard polarity subjectivity lexicon is applied to the emails. In practice, you should adjust the lexicon for common terms in emails that can skew results. For example, people often start correspondence with "Dear" and end with "Sincerely." These terms appear in the standard polarity dictionary. A previous chapter covers how to change the polarity table in qdap called `key.pol`. If needed, refer to Chapter 4 to adjust specific terms in your analysis.

Calculating polarity can be time-consuming on large corpora. The progress bar list apply, `pblapply`, function is once again helpful as it updates the calculation output.

```
three.ent.polarity<-pblapply(
three.ent.emails,polarity)
```

The `three.ent.polarity` object is a list containing the original text and data frames with polarity information for each email. Using `pluck` and data frame name, select individual polarity data from each list element. The object is still a list, but no longer contains the original text. The polarity information is now a list of data frames for each email. Each of those data frames contains one row and six columns. The average polarity score for an email will become the distribution of the box plot. Within the next line, the apply function is passed double brackets and the name of the specific column "ave.polarity." Since the data frames contain a single row, the single polarity value is selected. The apply function is nested inside `unlist` to convert the average polarity values to a numeric vector. This vector is the basis for the box plot distributions.

```
three.ent.polarity<-pblapply(
three.ent.emails,polarity)
score.list<-pluck(three.ent.polarity,"group")
scores<-unlist(pblapply(score.list, "[[",
'ave.polarity'))
```

Next organize the email polarity scores into a data frame to make the plot. The first column of the new data frame is the numeric vector "scores." The second column, "group," will contain the entity information. To this point, the code has been executed in sequence for all emails. The data needs to be correctly re-associated to an entity. As a result, the "group" column will have three different character strings corresponding to "Senate," "White House" and "Russia." The `rep` function repeats a value or string a specific number of times. In the code below, the string "Senate" is repeated according to the length of the grep object "senate." Next "White House" is repeated 109 times referencing the `white.house` object length. Lastly the same is done for "Russia." In your own analysis be sure to change the repeated value to the correct entity name and perform the repetitions in the same order as the script executes. For instance, if the code repeated "Russia" first, the polarity distributions would be inaccurate. Another best practice is to use the length function instead of a specific integer. This ensures the correct number of repeated values as you change the code for other entities.

```
scores.df<-data.frame(score=scores,
                group=c(rep('Senate',length(senate)),
```

```
rep('White House', length(white.house)),
rep("Russia",length(russia))))
```

The last step is to pass the two column data frame to ggplot. Add the x, y variables and define how the data is grouped in the first line. In this case, the "group" vector contains three distinct values corresponding to entities. Once the data is loaded, use geom_boxplot along with fill=group to construct the first layer of the graphic. The next layer calls a predefined palette mimicking Google Docs. In your own work, you can specify other themes or explicitly define palettes. The third layer directs ggplot to "jitter" the points. Jittering means to move the dots slightly from the exact y value. In this visualization, each point on the plot represents an email. Within an entity class, the polarity scores could be the same for different emails. Jitter ensures that the points do not completely overlap so the audience can distinguish each point's polarity score. The last line adds a title to the box and whisker plot in Figure 8.8.

```
ggplot(scores.df, aes(x=group, y=score, group=group)) +
  geom_boxplot(aes(fill=group)) +theme_gdocs() +
  geom_jitter(colour="gray40",width=0.2,alpha=0.3) +
  ggtitle("NER Polarity")
```

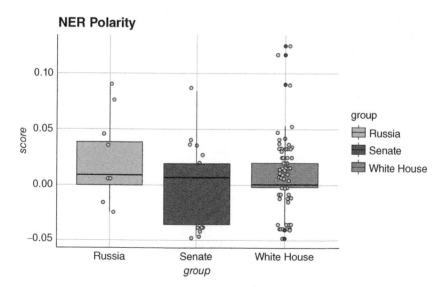

Figure 8.8 A box and whisker plot of Hillary Clinton emails containing "Russia," "Senate" and "White House."

Interestingly, one is led to believe that emails mentioning the Senate and White House contain less positive language than emails mentioning Russia. Upon deeper examination, two distinct negative clusters appear to skew the distribution for both the Senate and White House emails. Despite this observation, the median values are similar among all entities. Finally, the jittered points help your audience understand that the number of White House emails is significantly higher than the other entities. This should inform any conclusions that can be drawn from this graphic.

8.4.4 Stock Charts for Entities

Stock charts are line charts plotted with the Y axis representing price and the X axis representing a unit of time. Stock chart views help traders understand historical prices and can illustrate a pricing trend. A stock chart graphs a share price as a time series. In fact, time series stock charts are so common that two lines of code below from the quantmod package create the basic plot in Figure 8.9.

```
library(quantmod)
getSymbols("msft",src="google")
plot(MSFT)
```

If the corpus is arranged chronologically, a similar time series visual can be constructed. Instead of a share price, the Y axis can be the average polarity for a document or another continuous document attribute line nchar. To capture the evolving polarity over time another column containing periodicity needs to be appended to the scores.df object. A text miner can add specific dates, or in this case a sequence of numbers for the X axis. Appending a number sequence is easier, but is not as precise as using dates. A simple number sequence treats each email as a standalone incident like a stock's closing price. Using dates can be more laborious to identify within raw text, but can be insightful to group by date to arrive at the average. Using the exact date can give you other

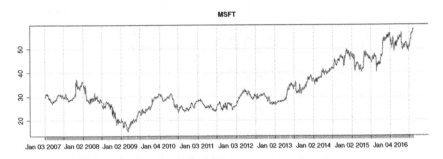

Figure 8.9 The Quantmod's simple line chart for Microsoft's stock price.

information analogous to an "opening" or "closing" share price. This example uses the seq function to append a number in sequence in the "chronological_ order" column. Since you are examining three entities you need to create three unique sequences matching to the number of emails per entity. The first 22 rows will have a sequence of numbers 1 to 22, then row 23 will start with 1 and progress to 109, and finally row 132 will start with another sequence for the remainder of the data frame.

```
scores.df<-data.frame(scores.df,
                  chronological_order=
                     c(seq(1:length(senate)),
                     seq(1:length(white.house)),
                     seq(1:length(russia))))
```

For easier comparison the code below creates three separate objects. Each object is a ggplot2 time series of the entity's polarity. Pass in scores.df with the periodicity information and index the exact rows for each entity. In this example, the Senate was mentioned 22 times, the White House 109 and Russia 12. For each subsequent plot be sure to advance the row index by 1 as shown. The rest of the code adds the line, uses the Google documents theme, removes the default legend and adds a title.

```
se.plot<-ggplot(data=scores.df[1:22,],
aes(x=chronological_order, y=score,
group=group, color=group)) +
geom_line() + theme_gdocs() +
theme(legend.position="none") +
ggtitle("Senate Polarity")

wh.plot<-ggplot(data=scores.df[23:131,],
aes(x=chronological_order, y=score,
group=group, color=group)) +
geom_line() + theme_gdocs() +
theme(legend.position="none") +
ggtitle("White House Polarity")

ru.plot<-ggplot(data=scores.df[132:143,],
aes(x=chronological_order, y=score,
group=group, color=group)) +
geom_line() + theme_gdocs() +
theme(legend.position="none") +
ggtitle("Russia Polarity")
```

The three objects can be plotted concisely using grid.arrange. This useful function from gridExtra lets you place multiple ggplot2 visualizations

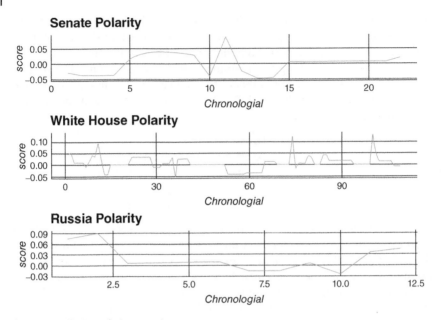

Figure 8.10 Entity polarity over time.

into a table. Within the function you can specify the number of rows and columns to place individual ggplots. However, the code below aligns the three time series into a single column to create Figure 8.10. The lengths between entities vary greatly so grid.arrange adjusts the X axis accordingly. To avoid any confusion, you may prefer to plot each time series separately.

```
grid.arrange(se.plot, wh.plot,ru.plot)
```

8.4.5   Reach an Insight or Conclusion About Hillary Clinton's Emails

This case study demonstrates named entity recognition methods on multiple emails. As a "journalist" you quickly identified people, places and organizations in the corpus. To a large extent, The Secretary of State mentions expected organizations. However, the organization data contains less frequent yet interesting entities. For example, "Egyptian Military Intelligence Service," "National Libyan Army" and "European Court of Human Rights" could all prove to be emails worthy of additional scrutiny.

Next, you plotted the identified locations on a map using Google's geocoding service. The map reveals the significant amount of time the Secretary wrote about Eastern US locations as well as Middle East affairs. A zoomed in map can provide more context as you explore the locations. For instance, the European map illustrates Secretary Clinton's multiple mentions within the Netherlands as opposed to no mentions in Poland.

Using the polarity function you can also begin to understand the author's attitude towards entities. The box and whisker plot illustrates the overall polarity distribution of multiple entities. The box plot lets you see the distribution compared to other entities by placing information side by side. In contrast the line charts represent changing language throughout time similar to a stock price chart. Using the line chart, an audience can gauge shifting attitudes towards an entity if the corpus is arranged chronologically.

8.6 Summary

In this chapter you applied various openNLP techniques to identify proper nouns such as people, locations and organizations from thousands of emails. Specifically you learned:

- what is the OpenNLP Project
- the basics of R's OpenNLP package
- an example of syntactic parsing
- what is named entity recognition (NER)
- to load NER libraries
- to perform NER
- to use an API to get latitude and longitude of recognized locations
- to create a bar chart of recognized organizations
- to use a heat map to understand the entity interactions
- to apply polarity scoring to individual entities in a box and whisker plot
- to chart document polarity over time for specific entities.

9

Text Sources

In this chapter, you'll learn

- adjustable components of a URL
- automatically constructing URLs for web scraping
- to web scrape a single page
- to web scrape an entire forum
- using an API within an R package to retrieve text
- to set up a connection with Twitter's API to retrieve tweets
- using R without a dedicated package to parse an API's JSON response
- to use the tm package plugin to retrieve a WebCorpus
- to use XML to parse an RSS feed
- simple methods for reading files types csv, .txt, MS Word and Excel documents
- to make a plain text document (.txt) from a PDF
- to use a Microsoft API to identify text in an image file
- two ways to read in multiple files in a folder.

9.1 Sourcing Text

When you are working on a text mining project, you may be confronted with various data sources. For example, if social media text is needed and an application program interface (API) exists then the tools you use to collect the text are straightforward. Other times, you may need to scrape online text which is essentially a custom script for each site due to the HTML variations. As a result, web scraping can be laborious, but it is necessary in some text mining projects. In other text mining pursuits, you may have text from organizational files such as Word documents or PDFs. The tools and approaches to organize text in each case vary. This chapter should serve as a reference guide if you get stumped in your own text mining efforts.

Text Mining in Practice with R, First Edition. Ted Kwartler.
© 2017 John Wiley & Sons Ltd. Published 2017 by John Wiley & Sons Ltd.

9.2 Web Sources

Online growth has created communication mediums including changing text usage. This section of the book encompasses common methods to obtain online text for analysis. Two methods include web scraping and parsing structured feeds like APIs.

When scraping it is ethical to strictly abide by the terms of service and a website's domain crawling restrictions. When consuming an API it is also important to follow the provider's terms. Web content providers are justified in wanting some control over their content while also avoiding overloading web servers. Thus, to avoid trouble, only scrape a web page or use a data feed after you have read and understood any restrictions.

There are many approaches to gathering online text, but this chapter focuses on the intuitive `rvest` package, APIs and extensible markup language (XML) parsing. Figure 9.1 illustrates the online text sourcing breakdown that is exemplified in this chapter

9.2.1 Web Scraping a Single Page with rvest

To begin, you will use Hadley Wickam's `rvest` package. Hadley Wickam created the R package to mimic the python package "Beautiful Soup," which collects information from web pages. The package's unusual name is because it is used to "ha-`rvest`" web pages.

The easiest manner to learn web scraping with this intuitive library is on a single web page. It is best to start by examining the page. In this example, it is an amazon.com help forum webpage found at the link below. http://www.amazon.com/gp/help/customer/forums/ref=cs_hc_g_tv?i.e.=UTF8&forumID=Fx1SKFFP8U1B6N5&cdThread=Tx3JJLVOS6N6YSD

Figure 9.2 is a partial screenshot of this forum discussion on Prime movies. For more information, navigate directly to the webpage.

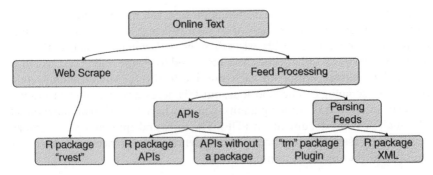

Figure 9.1 The methodology breakdown for obtaining text exemplified in this chapter.

Figure 9.2 An Amazon help forum thread mentioning Prime movies.

Like all web pages there are many elements. As a text miner you are probably only interested in the forum thread, authors and dates. So care must be taken to avoid collecting multitudes of unwanted information such as extraneous links and header text. For example, the right-hand side header "Frequently asked questions" is a piece of text that rarely changes in the forum, so likely not the text you will want to extract.

The first step in scraping the post information is to load the `rvest` library, specify the web address and read the html source code of the webpage. The code below demonstrates this basic first step by creating the `url` object and passing it to the `read_html` function. The `url` object is the text string between quotes representing the web address.

```
library(rvest)
url<-'http://www.amazon.com/gp/help/customer/forums/
ref=cs_hc_g_tv?i.e.=UTF8&forumID=Fx1SKFFP8U1B6N5&cdThr
ead=Tx3JJLVOS6N6YSD'
page<-read_html(url)
```

The single forum page has been captured in your R session in the `page` object. It contains all elements of the page so the interesting text must be extracted. An easy method is to select the cascading style sheet's (CSS) class for the exact element needed. However, as an R programmer, you may not be able to read CSS or HTML code. Luckily, there is an easy Chrome extension called "SelectorGadget." It is available at www.selectorgadget.com. Once it is installed, you will notice a small magnifying glass in the upper right section of your Chrome browser. To identify the CSS section of an HTML document, you press the magnifying glass icon and a popup will appear. As you navigate the webpage, shared CSS classes become highlighted. Once you have the item you want to scrape highlighted, select it by pressing your mouse button. The popup

Figure 9.3 A portion of the Amazon forum page with SelectorGadget turned on and the thread text highlighted.

window will display the highlighted section's CSS class. Figure 9.3 shows the previous web page with the text thread highlighted and the popup showing a CSS class called ".thread-body."

Armed with the CSS class knowledge, extracting the thread text is straight-forward. The code below passes in the page and a string of the CSS class to easily extract sections of HTML code. In this example, you use the ".thread-body" that the SelectorGadget extension identified. Once the CSS element has been identified, the last line of code only extracts the text.

```
posts<-html_nodes(page, '.thread-body')
forum.posts<-html_text(posts)
```

The forum.posts object is a character vector and has correctly scraped the four posts on this page, ignoring all the other text. In isolation, it may not be useful, so you may have to extract other text from the page. In this next code section, you switch from CSS class to using Xpath. The rvest package allows you to pass in either a CSS or Xpath syntax to scrape HTML elements. Xpath is a syntax to identify sections within extensible markup language (XML). Xpath navigates XML documents, identifying specific nodes and in the context of web scraping can be used to save the specific text elements. Although Xpath is out of scope for this book, it is used here as a demonstration with basic explanation.

Specifically, XPath is used in the first code line. Instead of a CSS class, the html_nodes function now has xpath="//a". This line identifies a web link anywhere in the scraped document. The double slashes represent any-where in the document. The "a" captures all links because links are defined with the HTML "<a>" tag. However, this will capture more than 100 links on the webpage. So the next line of code uses grep to identify the index position of any link's text matching "Permalink." You know that the thread links have this text because you can see it above each forum post. The grep function uses regular expression wildcards, asterisks, for pattern matching. This lets grep

Table 9.1 Forum.posts and thread.urls using `rvest`.

Forum posts	Permalink
"Why,after paying for Amazon Prime,are there so many movies and things like that,that I still have to buy or rent instead of being able to watch them for free?"	www.amazon.com/gp/help/customer/forums/ ref=cs_hc_g_pl?i.e.=UTF8&forumID= Fx1SKFFP8U1B6N5&cdThread=Tx3JJLVOS6N6YSD &cdPage=1&cdMsgId=Mx34N1XDE8MHZGL #Mx34N1XDE8MHZGL
"Because the studios don't want their videos in the Prime program."	www.amazon.com/gp/help/customer/forums/ ref=cs_hc_g_pl?i.e.=UTF8&forumID= Fx1SKFFP8U1B6N5&cdThread=Tx3JJLVOS6N6YSD &cdPage=1&cdMsgId=Mx3MZ3Z2ETIH48T #Mx3MZ3Z2ETIH48T
"Still don't understand why have to pay for movies with Amazon Prime."	www.amazon.com/gp/help/customer/forums/ ref=cs_hc_g_pl?i.e.=UTF8&forumID= Fx1SKFFP8U1B6N5&cdThread=Tx3JJLVOS6N6YSD &cdPage=1&cdMsgId=Mx4K5HKP7XZ2R1 #Mx4K5HKP7XZ2R1
"Because not all movies are part of the Prime subscription. You were advised, when you signed up for Prime, that Prime allowed you to view eligible movies for free. Eligible, as in the rightsholder agreed to be part of Prime free movies. Studios still want to make money on their product. How hard is that to understand? Do you work for free?"	www.amazon.com/gp/help/customer/forums/ ref=cs_hc_g_pl?i.e.=UTF8&forumID= Fx1SKFFP8U1B6N5&cdThread=Tx3JJLVOS6N6YSD &cdPage=1&cdMsgId=Mx3KBBGVAS0BC6X #Mx3KBBGVAS0BC6X

look anywhere in the links text. In this example, links 19, 26, 34 and 42 are "Permalinks" among all 150. The next line of code selects the attributes of the links. The 150 links are subset by adding the bracketed `thread.urls` after selecting the "href" attributes, thereby capturing the Permalinks. In HTML, the "<href>" tag is the destination address of the post. These links can be used as unique identifiers because each forum post is assigned a specific address. The last line of code adds "amazon.com" to the front of the scraped link addresses, so they are complete. Table 9.1 indicates the four "Permalinks" alongside the specific forum post.

```
links<-html nodes(page,xpath='//a')
thread.urls<-grep("*Permalink*", links)
thread.urls<-html_attr(links,"href")[thread.urls]
thread.urls<-paste0('www.amazon.com',thread.urls)
```

Next, you may want to get the author name for each post. When you investigate the page, you should note that each author's name is actually a link to their profile page. An example public profile is here:

http://www.amazon.com/gp/profile/AD43H667F2BGW

Notice in the middle of the link is the word "profile." To obtain profile link information, change the previous grep code to search for "profile" within all the links you already captured. However, instead of getting the HTML attributes, the author name is needed. To get author names, pass in the links, indexed by "profile", into "html_text." The "authors" object extracts the author's screen-name, since the name is the text associated with each link. You may also want to get the links themselves. If so, you now extract the link attributes using html_attr and append www.amazon.com.

```
profile<-grep("*profile*", links)
authors<-html_text(links[profile])
author.profiles<-html_attr(links,"href")[profile]
author.profiles<-paste0('www.amazon.com',author.
profiles)
```

The authors object is a character vector with the names "Greg Perkins," "**MeyaPaloozza**," "Carol" and "Artist." The author.profiles object contains another character vector with respective profile URLs.

Lastly, the objects should be organized into a succinct data frame. This is done with the following code.

```
final.df<-
data.frame(forum_post=forum.posts,author=authors,author_
urls=author.profiles,thread_urls=thread.urls)
```

9.2.2 Web Scraping Multiple Pages with rvest

The previous section may have seemed like a lot of work, but do not be overwhelmed. This section creates functions to automate over the entire forum rather than specifying individual pages. It is best to start with a single page so you can learn the page structure and identify the correct elements. Once you understand a single page you can write functions for iterating the web scraping process over hundreds of pages.

To scrape more than one page at a time, you load the rvest, pbapply and data.tables packages. In the previous section you had exposure to rvest. The pbapply library adds progress bars to R's set of apply functions. Since you are scraping many hundreds of pages it is useful to visually understand the progress. The last package, data.table, extends the data frame functionality to work in a faster and more efficient manner. The data.table package has a particularly useful function rbindlist used to organize the scraped pages.

```
library(rvest)
library(pbapply)
library(data.table)
```

After loading the packages, you should explore the first couple pages of the forum. In the link below, you can see `forumID=` and `cdPage=`. The ID parameter is the unique identifier to the specific forum. Amazon's general retail forum ID is "Fx1SKFFP8U1B6N5." A similar forum is devoted to Kindle products and has a different ID, "Fx1FI6JDSFEQQ7V." As you explore the page, you can see 40 individual pages of forum topics among other links. If you change the `cdPage=2` number, your browser will load the corresponding page pertaining to the unique forumID. Figure 9.4 is a screenshot of Amazon's general help forum.

http://www.amazon.com/gp/help/customer/forums/ref=cs_hc_g_pg_pg40?i.e.=UTF8&forumID=Fx1SKFFP8U1B6N5&cdPage=2

In order to scrape the entire forum, you must first scrape the 25 links in the "Recent questions" section for each of the 40 pages. After that, you can scrape the actual conversations. As a result, scraping the specific conversations is a two-step process. The two-step workflow is presented in Figure 9.5.

From the original web address, you first construct links representing all 40 pages. In the code below, you use the `paste0` command with the base for each address. The page parameter is replaced with a sequence of numbers from 1 to 40 using the `seq` function. The forum URLs for all 40 pages of the forum are now a single character vector.

```
forum.urls<-paste0('http://www.amazon.com/gp/help/cus-
tomer/forums/ref=cs_hc_g_pg_pg40?i.e.=UTF8&forumID=Fx1
SKFFP8U1B6N5&cdPage=',seq(1:40))
```

Figure 9.4 A portion of Amazon's general help forum.

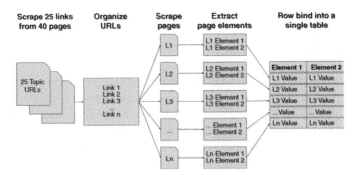

Figure 9.5 The forum scraping workflow showing two steps requiring scraping information.

Each of the 40 pages contains 25 conversation threads. To automate the collection process, create a custom function searching for the conversation thread links. The `url.get` function below uses similar syntax from the last section, but it is a function. The function is passed "x" which is a web URL. Then `read_html` reads that individual page. Next the html nodes are examined using `Xpath`, which will identify any table called "thread-list-table" and then cells in the table containing links. The `SelectorGadget` extension can be used to identify the table name for entering in the code below. Then the `Xpath` identifies any cell ("//" and the html tag "<td>") that contains the "<a>" and "<href>" link tags. With this brief explanation, the `Xpath` syntax becomes identifiable. Once the "thread-list-table" cells with links are recognized, the text is extracted using `html_text`, which is a partial address so "http://www. amazon.com" is pasted to the link text as the function's output.

```
url.get<-function(x){
  page<-read_html(x)
  links<-html_nodes(page,xpath=
                    "//table[@class='a-bordered
thread-list-table']//td/a/@href")
  links<-html_text(links)
  links<-paste0('http://www.amazon.com',links)
}
```

To obtain all conversations in the forum, the `url.get` function is applied to each of the 40 previously constructed URLs. The function is applied using `pblapply` and then unlisted to get a vector of all conversation links in the entire forum. You could use `lapply` but the progress bar version is more informative. After running the code below, you will have a vector of 1000 forum conversation links. This is because 25 links were identified for each of the 40 pages (25*40).

```
all.urls<-pblapply(forum.urls,url.get)
all.urls<-unlist(all.urls)
```

The next bit of code is another custom function. It is very similar to the previous single web scrape code. The single page workflow is followed exactly, within the forum.scrape function. The function is passed one of the 1000 conversation URLs as "x". The individual URL is read, and nodes for the ".thread-body" are identified. Next the html_text function extracts the post threads. Once again, link nodes are found using Xpath with "//a." Then, URLs for each Permalink, author and author profile are extracted. Consult the previous section, which describes the elements in more detail if you need a full explanation of grep, html_attr or html_text. The extracted elements of each page are organized into a small data.table representing the entire conversation on the URL. The data.table function was used instead of data.frame because it is more memory efficient. The output of the function is a data table with the information for the specific conversation thread.

```
forum.scrape<-function(x){
  page<-read_html(x)
  posts<-html_nodes(page, '.thread-body')
  forum.posts<-html_text(posts)
  links<-html_nodes(page,xpath='//a')
  thread.urls<-grep("*Permalink*", links)
  thread.urls<-html_attr(links,"href")[thread.urls]
  thread.urls<-paste0('amazon.com',thread.urls)
  profile<-grep("*profile*", links)
  authors<-html_text(links[profile])
  author.profiles<-html_attr(links,"href")[profile]
  author.profiles<-paste0('amazon.com',author.
profiles)
  final.df<-data.table(authors,author.profiles,forum.
posts,thread.urls)
  return(final.df)
  }
```

The forum scrape function is applied to the 1000 conversation URLs. Again, pblapply is used to construct a large list with 1000 elements. Each element is a data table representing a single page's conversation specifics including text, author info and links. Figure 9.6 is a visual representation of the data after running the code. In this structure, each data table could have a different number of rows depending on the number of thread posts. However, each data table has the same columns.

```
amzn.forum<-pblapply(all.urls,forum.scrape)
```

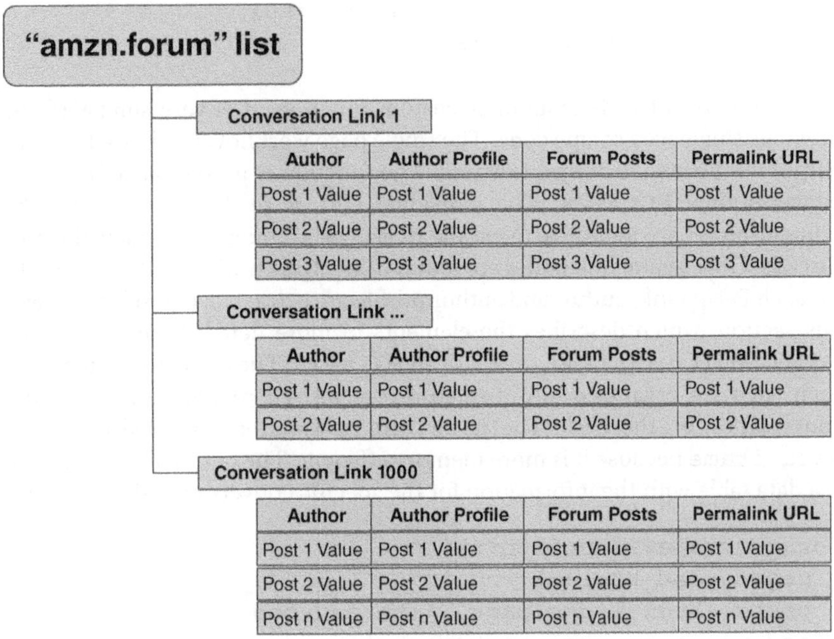

Figure 9.6 A visual representation of the amzn.forum list.

The last step is to row bind the entire list into a single table. The `data.table` package function `rbindlist` quickly performs row binds on large lists. Once it is organized into a table you can begin your cleaning and text mining analysis. If needed, the Permalink value contains unique message and conversation identifiers. Figure 9.7 illustrates the effect of "`rbindlist`" on the amzn.forum list.

```
amzn.forum<-rbindlist(amzn.forum, use.names=TRUE)
```

Most forums are constructed with different link and HTML structures. The following code applies the same methodologies to scraping a new forum structure. Instead of a customer service forum, this forum is a bitcoin general discussion. An interesting analysis may be predicting daily bitcoin prices based on the text in this forum. In this example, the link construction and nodes are specific to the forum so the Amazon support forum code would not work. Executing this code may seem redundant, but will demonstrate the changing function parameters and introduce the pipe (%>%) operator.

Load `rvest` and `pbapply`.

```
library(rvest)
library(pbapply)
```

"amzn.forum" data table

Author	Author Profile	Forum Posts	Permalink URL
Post 1 Value	Post 1 Value	Post 1 Value	Post 1 Value
Post 2 Value	Post 2 Value	Post 2 Value	Post 2 Value
Post 3 Value	Post 3 Value	Post 3 Value	Post 3 Value
Post 1 Value	Post 1 Value	Post 1 Value	Post 1 Value
Post 2 Value	Post 2 Value	Post 2 Value	Post 2 Value
Post 1 Value	Post 1 Value	Post 1 Value	Post 1 Value
Post 2 Value	Post 2 Value	Post 2 Value	Post 2 Value
Post n Value	Post n Value	Post n Value	Post n Value

Figure 9.7 The row bound list resulting in a data table.

The following link will direct your browser to a page in the general discussion forum at bitcointalk.org.

https://bitcointalk.org/index.php?topic=976903.4460

Notice the topic is "976903.4460" in the link. If you navigate to the next discussion page, the address changes slightly. In the new link below, you should notice that the last four digits have changed. It turns out that this forum has a topic number identifier followed by the post count. Each page loads 20 messages of the forum so the "4460" has changed to "4480."

https://bitcointalk.org/index.php?topic=976903.4480

The code below constructs all web addresses for the specific topic. The paste0 function is used to append the base of the link to a sequence of numbers. The number sequence increases by 20 instead of 1 because by= is a sequence parameter. The sequence runs from 0 to 5040 but should be updated to a larger number as the conversation increases over time.

```
all.urls<-paste0('https://bitcointalk.org/index.php?to
pic=976903.',seq(0,5040,by=20)
```

In this example, 253 unique web addresses were created. The next code section creates a custom function called "forum.scrape." One of the forum addresses is passed to the function. It is read using read_html as before. However, this function adds "Sys.sleep(1)." Interjecting this code into a function forces R to wait a full second before proceeding. As you iterate over hundreds of web pages this may keep your function from overloading servers or being blocked. It may not be necessary or may need to be increased, depending on the web site you are working on. The next line of code should look familiar but in a different, unaccustomed structure. The object "posts" is created by finding `html_nodes` with CSS class ".post." In the previous example, the CSS class was ".thread-body." Then the text is extracted. This code line uses the "%>%" operator instead of having multiple code lines. The "%>%" is called the pipe operator, which simply forwards the object into another expression or function. While a bit more confusing than the previous examples, the code is more compact. The "%>%" moves the object "x" in sequence to `html_nodes`, `html_text` and then `as.character` rather than having three separate lines. Lastly, the posts for each discussion are collapsed into a single large text using paste with collapse.

```
forum.scrape<-function(forum.url){
  x<-read_html(forum.url)
  Sys.sleep(1)
  posts<-x%>%html_nodes(".post")%>%html_text()%>%as.
character()
  posts<-paste(posts, collapse='')
  return(posts)
}
```

Next the forum.scrape function is applied to all 253 bitcoin forum posts. Once again, the code employs the progress bar version of `lapply`. This applies the `forum.scrape` function in sequence from 1 to 253. Your R console will pause for one second for each page because the custom function contains a system sleep.

```
all.posts<-pblapply(all.urls,forum.scrape)
```

Lastly, the base R `do.call` with `rbind` is applied to the all.posts list. The result is a matrix with 253 rows and one text column. Each row represents a page of the forum with 20 posts collapsed into a single block of text.

```
bitcoin<-do.call(rbind,all.posts)
```

The text in this example may be too raw for meaningful exploration. You will need to improve on the custom function to capture authors and dates if you want to attempt to predict bitcoin prices. However, this example does introduce useful new concepts. It illustrate not only changing a CSS class, but also system sleep, the pipe operator and using `do.call` for binding a list.

9.2.3 Application Program Interfaces (APIs)

Application program interfaces (APIs) are the links between applications allowing information to be shared. An API call is similar to calling the telephone operator for information. You need to dial a specific number to reach the operator. If you dial incorrectly you will not get a response. Once you reach the operator, you have be specific and structured in your information request. The operator is unable to help you if your intent is not clear. If the operator understands the request, he or she responds with the information you need and then you can hang up. An API works in an analogous manner.

When performing an API call, your R console will connect to another application. Your script will request information from the remote server. If the request is understood and within the API rate limits, the service will return some information that can be saved or analyzed in R. You will need to be connected to the Internet to make the API calls presented in this section. APIs represent an efficient way to gather text or other data. API data is structured and usually has other important metadata with it. As a result, APIs are a great way to access new text for analysis. To start, this section demonstrates R packages devoted to specific APIs. Next, you will learn how to make an API call without a dedicated package using a package called jsonlite. Lastly you will learn ways to consume standard web feeds like a Rich Site Summary (RSS) using XML.

9.2.4 Newspaper Articles from the Guardian Newspaper

The Guardian newspaper is an internationally recognized news organization. The newspaper is headquartered in the United Kingdom with satellite offices around the world. The newspaper has a long history of balanced and independent reporting dating back to 1821. The newspaper's profits do not benefit a media tycoon or shareholders. Instead the profits are redirected to support journalism within the Scott Trust Limited company. The trust company was formed to ensure the newspaper's longevity and has an investment fund supporting the newspaper's annual losses.

The newspaper offers an API for developers. Through the API you can access over 1.5 million pieces of newspaper content. According the API website, the newspaper hopes to shape the future of digital journalism by offering the service. To that end, the API's blog explains various topics such as cloud computing or building a chat bot.

To access the API you need to obtain a developer key. A key is a long, cryptic character string that is sent along with your request to the newspaper's servers. This helps the newspaper track your specific usage and ensure that you are not abusing the system. You should keep your key private. Register for an API key with the link below or search online for "Guardian Newspaper API." Submit the form to receive a key in your registered email inbox.

https://bonobo.capi.gutools.co.uk/register/developer

The rest of the Guardian API code assumes that you have obtained a developer key. First, load the GuardianR and qdap packages. The GuardianR package wraps API calls into simple R functions. At the time of writing, the package version is 0.6. In subsequent versions, the code below may need to be changed slightly. The "key" object below is the character string where you should paste in your unique key. The get_guardian function searches for the term "Brexit" within a date range. You can change the search term and dates within the function to get different articles. The last parameter within the get_guardian function passes in your API key. Keep in mind APIs change often so the code may vary.

```
library(GuardianR)
library(qdap)
key<-'xxxxxxxx-xxxx-xxxx-xxxx-xxxxxxxxxxxx'
text <- get_guardian("Brexit", from.date="2016-07-
01",to.date="2016-07-06",api.key=key)
```

At the time of writing, the API will return 208 Guardian articles mentioning "Brexit" when you execute this code (with a proper API key). For context, "Brexit" was the name of the United Kingdom referendum to exit the European Union. In Britain, there was an impassioned debate leading up to the public vote, so many articles were written on the subject.

The "text" object is created as a data frame. This makes retrieving text straightforward because the function parses the API response, specifically the JavaScript object notation (JSON), for all 208 articles into a manageable data frame. The API response data frame has 208 rows with 27 columns. Table 9.2

Table 9.2 The Guardian API response data for the "text" object.

API data	Continued API data
id	Standfirst
sectionID	shortUrl
sectionName	Wordcount
webPublicationDate	Commentable
webTitle	allowUgc
webURL	isPremoderated
apiURL	Byline
newspaperPageNumber	publication
trailText	newspaperEditionDate
seadline	shouldHideAdverts
showInRelatedContent	liveBloggingNow
lastModified	commentCloseDate
hasStoryPackage	Body
score	

lists the returned API data names. The upcoming code will focus on the `text$body` and `text$id`. If you need other article information for your analysis, consult the API documentation for attribute definitions.

The `text$body` vector contains the article text. In order to analyze the article, you will need to change its encoding, remove any embedded links and eliminate HTML tags. The `iconv` function performs international character string conversion. The API sends text to your R console as "latin1," but R text mining is easier in "ASCII." The `iconv` function below converts the `text$body` to create the "body" object. `Gsub` is used next on the "body" object along with a regular expression identifying "http." Once http is found, an empty character is substituted in its place. In effect, this will remove all hyperlinks. Lastly, you apply `bracketX` to remove any HTML tags within "<" and ">." As written, the `bracketX` function will remove text in between *any* brackets. The last line of code creates a simple data frame of clean article text with a unique identifier. With basic cleaned text, "text.body" data frame can be the basis of your text mining project.

```
body<-iconv(text$body, "latin1", "ASCII", sub="")
body<-gsub('http\\S+\\s*', '', body)
body<-bracketX(body, bracket="all")
text.body<-data.frame(id=text$id,text=body)
```

9.2.5 Tweets Using the twitteR Package

Twitter is a micro-blogging service letting people stream information in individual feeds. Millions of Twitter users send out 140 character "tweets" about any subject, organized by screen name and chronologically. Users sign up to influential, interesting or celebrity feeds, and in turn can tweet, or retweet, information in 140 character messages. Most Twitter feeds are public, so the worldwide service is a great way to get social media text that is timely. The free API access is not the "full fire hose" of tweets, but it does provide enough text to make an interesting text mining project. Be aware that Twitter users tweet profanity, misspell words and use abbreviations, so the text can be challenging to clean. R has multiple packages to access the Twitter API. The simplest one is called "twitteR." Once again remember APIs change so the code may vary.

To get developer access, you must be a Twitter user. First sign up for the Twitter service as a consumer at www.Twitter.com. Then sign in and navigate to the developer link below to create an application. On this page, there is a "Create New App" button that navigates to a form. Fill out the form, accept the terms and click "Create Your Twitter App."

https://apps.Twitter.com/

Once it is complete you will be taken to the application's settings page. Click the "Keys and Access Tokens" link. For Twitter's API, you need to get a consumer key and secret consumer key. Both are used for authentication in your R console. After loading the twitteR library, copy and paste both keys into the code snippet below to authenticate your R session. The last line performs "Open Authorization" (OAuth) between your R console and the Twitter API service. The OAuth protocol is an open standard for identity authorization without exposing passwords. The protocol is key-based and used by many services such as LinkedIn and Google.

```
library(twitteR)
consumer.key <- 'long_string_of_characters_from_
Twitter'
consumer.secret <- 'another_long_string_of_charac-
ters_from_Twitter'
setup_Twitter_oauth(consumer.key, consumer.secret)
```

The first time you run this code you will be prompted to save your authorization credentials, and a browser will open to confirm permissions. If you decide to save the small authorization credential file to your working directory, the browser pop-up will only occur once.

Once you are authenticated, you have access to a range of interesting API calls. A simple way to find tweets is to use a search term. The searchTwitter function is passed a term, the number of tweets you are requesting and the language. Here the tweets.one object is searching Twitter for 2000 AmazonHelp tweets in English. Previously the get_guardian function called the service and automatically organized the response into a data frame. The searchTwitter function does not do this. As a result, another function twListToDF creates a data frame from the search results list. The tweets. one.df object contains 2000 tweets mentioning AmazonHelp. This does not mean that the tweets were made by AmazonHelp. If another user mentions or replies to AmazonHelp, the tweets may show up in the search results. If you need to capture only tweets from AmazonHelp, you can subset the data frame by screen name equal to "AmazonHelp." In doing so, the tweets data frame is reduced to only tweets authored by "AmazonHelp."

```
tweets.one<-searchTwitter("AmazonHelp",
n=2000,lang='en')
tweets.one.df<-twListToDF(tweets.one)
tweets.one.df<-subset(tweets.one.df,tweets.one.
df$screenName=="AmazonHelp")
```

An interesting Twitter phenomenon is the use of the "#" or hashtag followed by a word. This succinctly conveys meaning, news trends or emotion in tweets. If you want to examine hashtag trends, you need to identify the most used hashtag at the time for a location.

Executing this code will print all worldwide trend location codes to your console. The codes are "Where On Earth ID" (WOEID) identifiers. These are reference codes for a specific location used by GeoPlanet and Yahoo developer APIs. The reference codes do not change, so they are used elsewhere in other APIs like Twitter.

```
availableTrendLocations()
```

Once you select a WOEID, you pass it to the `getTrends` function. Here the worldwide code "1" is used. This returns the top 50 trends in a data frame at the time of the API call for the entire world.

```
world.wide.trends<-getTrends(1)
```

The data frame containing worldwide trends information can be examined using this code. You can continue to change the first index number to examine the top hashtags or save the entire data frame.

```
world.wide.trends[1,1:4]
world.wide.trends[2,1:4]
```

In this example, the getTrends function is passed the USA WOEID number. To review all hashtags without other information, use the next code line. This will print all 50 top hash tags in the USA at the time of your API call.

```
usa.trends<-getTrends(23424977)
usa.trends[,1]
```

At the time of this writing the top trend in the USA was "#PowerPremiere." Once the trending hashtag is identified, you can pass a hashtag like "#PowerPremiere" into the `searchTwitter` function to get tweets about the topic. This can be particularly useful during a political crisis or major sporting event.

Another way to search for particular tweets is by user timeline. If you want to follow a specific user, celebrity or news feed, use the `userTimeline` function. Simply pass in the Twitter handle of the user and the number of tweets within `userTimeline`. This code is requesting 1000 tweets from the Twitter handle "datacamp." Once again, the `twListToDF` function changes the returned list to a data frame. The free API access will limit the number of tweets you receive in this API call. Thus, you may not get all 1000 tweets.

```
datacamp.com<-userTimeline('datacamp',n=1000)
datacamp.df<-twListToDF(datacamp.com)
```

9.2.6 Calling an API Without a Dedicated R Package

So far, you have learned how to use dedicated R packages to access APIs. However, there are thousands of public APIs that you may want to use to gather information. There are some useful packages that more broadly let you consume APIs. Generally, modern APIs come in two response formats, JSON and XML. A popular JSON package is jsonlite. For XML responses, you can use the XML package. Mastering these packages will help you consume data outside of the dedicated API R packages.

9.2.7 Using Jsonlite to Access the New York Times

In addition to the Guardian newspaper, the New York Times offers an API giving you access to articles. The New York Times was first published in 1851. The newspaper covers worldwide events from its headquarters in the United States. The newspaper's API endpoints can be directly accessed from a web address showing the raw JSON text. Sometimes, the most challenging part of using an API is understanding the direct JSON web address structure. Perform an online search for New York Times API to find the developer webpage. After signing up for a developer key, you can access a direct API response using the link below or review the API's documentation.

For the New York Times API, the address below will search for articles containing "United States." Notice how the terms are separated by "+." In addition to the search terms, you need to pass in the API key after "key=." Replace the "x" characters with your own developer key.

http://api.nytimes.com/svc/search/v2/articlesearch.json?q=united+states&api-key=xxxxxxxxxxxxxxxxxxxxxxxxxxxxxxxxxxxx

If you navigate to this page, your browser will load the JSON text. Information encoded as JSON text is a lightweight way to pass information to your machine without using significant bandwidth. However, it is difficult for a human to read this information. The text must be parsed from JSON to an organized state like a text vector or data frame. The jsonlite package is a popular R package to change the text to a more useable form. First, load jsonlite and then change the api.key object string to your developer key. The code below uses a single search term, "brexit," to create the search. term object. You can change the string to any other interesting character pattern. If needed, the prior link exemplifies the search structure for more than one term. The last line uses paste0 to construct the web address for the API response.

```
library(jsonlite)
api.key<-'long_string_of_characters_from_NYT'
search.term<-'brexit'
url<-
paste0('http://api.nytimes.com/svc/search/v2/article-
search.json?q=',search.term,'&api-key=',api.key)
```

If needed, use the following code to add a time constraint to your API request. The web address is similar but adds date parameters before closing with the API key. The date format is year, month and then day.

```
start.date<-'20160701'
end.date<-'20160706'
url<-paste0('http://api.nytimes.com/svc/search/
v2/articlesearch.json?q=',search.term,'&begin_
date=',start.date,'&end_date=',end.date,'&api-
key=',api.key)
```

Next you invoke the fromJSON function by passing in either of the constructed web addresses. Your R session will navigate to the web page and read the lines of text. The JSON text is presented in sequence. This means that the first article and its metadata are presented, then the next article and so on. The sequential API responses are organized into a list.

```
ny.times<-fromJSON(url)
```

The multipart API response, now a list, can be examined using specific element names. Some interesting API responses are shown individually below. For example, the web page, publication date and headline elements are named list elements. As a reminder, when working with a complicated object like this list response, you can access the object structure using str (ny.times). The names are nested in two list elements. Using names (ny.times$response$docs) will print the article metadata and content names.

```
ny.times$response$docs$web_url
ny.times$response$docs$pub_date
ny.times$response$docs$headline$main
ny.times$response$docs$snippet
ny.times$response$docs$lead_paragraph
ny.times$response$docs$abstract
```

If you want to access other New York Times API endpoints, you can review the API documentation. The structure and version can change so it is best to check the documentation if you get an error. The API response list is organized into a data frame for consistency with other chapter examples. The list index position is used to construct the data frame instead of element names.

```
ny.df<-data.frame(url=ny.times$response$docs[1],
          date=ny.times$response$docs[11],
          headline=ny.times$response$docs[9][[1]]
[[1]],
          snippet=ny.times$response$docs[2],
```

```
lead_para=ny.times$response$docs[3],
abstract=ny.times$response$docs[4])
```

9.2.8 Using RCurl and XML to Parse Google Newsfeeds

Next you may encounter extensible markup language (XML) based APIs. As with JSON based API responses, the text is presented in a lightweight text-only HTML page. The text's attributes are presented in a tree structure with parent and children nodes. This structure is similar to desktop file systems. The following example is not technically an API. Instead, it uses XML to consume a Google News RSS feed. Consuming an RSS was chosen because JSON is more popular, and an RSS feed does not require an API key making the code straightforward.

First, you load a new library RCurl, which is used for hypertext transfer protocol (HTTP) requests. The RCurl package allows you to generate web server requests, such as posting to a form and processing results from a web server. In this case, RCurl downloads the information from the XML structured web page.

Next, load "XML" into your R session. The XML library provides tools for parsing and creating XML structured information. After that, the qdap package is used for some additional text cleaning. Using qdap applies to this example and may not be needed for other XML text processing. Similar to the JSON example, you are creating a search term to construct a specific web address.

```
library(RCurl)
library(XML)
library(qdap)
search.term<-'Amazon+echo'
```

Here is the link you will be downloading to get XML information.
https://news.google.com/news?hl=en&q=Amazon+echo&i.e.=utf-8&num=100&output=rss

Investigating the link, you should notice the "q=" followed by the search term(s). The search term has two patterns "Amazon" and "echo." Next the "utf-8" specifies the character encoding for the response. The maximum number of returned google news articles is "100" for this type of query. The last part of the link asks the google server to return the information in an "RSS" format. Without this specification, the response will be a typical web page you would see in your browser. Figure 9.8 is a screen capture of the non-RSS Google News webpage.

In contrast, the RSS version presents only text from the news articles. It removes the left-hand links and images. Using the RSS styled link, you use the paste0 function to construct the final URL for parsing. The url object is the web address representing the Google News feed for articles mentioning Amazon and Echo. It is a best practice to copy the URL into a browser to ensure

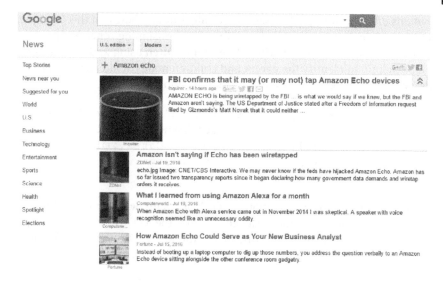

Figure 9.8 A typical Google News feed for Amazon Echo.

that the link is constructed correctly. If it is not, your browser will not be able to load any information.

```
url<-paste0('https://news.google.
com/news?hl=en&q=',search.
term,'&i.e.=utf-8&num=100&output=rss')
```

Next, you use the getURL command from RCurl. This makes a request to the server and allows your R session to download the text on the page. Since the RSS feed has an XML structure the next line is needed to parse it. The tree nodes of the XML document are contained in the document itself, so you have to pass in another parameter "useInternalNodes=T." At this point, the "rss" object is a XML document organized as a tree containing news articles. You still need to extract the parts of the tree based on node names to make the data useful.

```
rss<-getURL(url)
rss<-xmlTreeParse(rss, useInternalNodes = TRUE)
```

The next code lines use XPath to extract the document values. In this tree structure, each article is a parent node called "item." So the code below is searching for any ("//") "item" node. When it encounters an "item" node it will extract child nodes such as "/title" or "/link." The xpathSapply function follows an easy-to-understand structure. First, pass in the XML document. Next, pass in the XPath and finally the criteria for the object. The rvest package contains different functions to save node values or attributes. In this example, you are only collecting XML values. If you are consuming other XML documents the parent node name will not be "item." Often using xmlRoot(rss) can be helpful to identify the parent and children nodes of interest. This function will print the XML nodes making identification easier.

```
headline<- xpathSApply(rss,'//item/title',xmlValue)
link <- xpathSApply(rss,'//item/link',xmlValue)
date <- xpathSApply(rss,'//item/pubDate',xmlValue)
description <- xpathSApply(rss,'//item/
description',xmlValue)
```

In this example, the "description" node XML value will retrieve HTML in addition to the text. A quick way to remove the HTML is to use qdap's bracketX. This is done before organizing the RSS feed into a data frame in the final line.

```
description<-bracketX(description)
text<-data.frame(headline,link,date,description)
```

9.2.9 The tm Library Web-Mining Plugin

Another package worth exploring is used in conjunction with the tm package. Specifically, this package consumes newsfeeds and automatically constructs VCorpus objects. The automatic link construction, web site parsing and corpus construction make integration into a text mining workflow efficient. However, this plugin can encounter issues, as the web page structures and limits change. With so much automation, there is little flexibility to overcome errors. Still the web mining plugin can quickly gather text from specific web pages, so it is worth exploring. If you encounter errors, you should attempt to use rvest or XML parsing to gather the information. Using either of these packages is more laborious, but gives you more flexibility to overcome errors.

To begin, you load both tm and tm.plugin.webmining. The tm. plugin.webmining library function structures are consistent with tm functions. The code below uses functions WebCorpus and GoogleFinanceSource. These functions correspond to VCorpus and VectorSource from tm. The plugin library has multiple source functions for different web services such as Yahoo News or Google Finance. The WebCorpus function gathers metadata and the main text for the specified source. Then it constructs a VCorpus that can be used in a typical text mining workflow.

Here, the source is a Google Financial Newsfeed for the stock ticker "amzn."

```
library(tm)
library(tm.plugin.webmining)
```

```
goog.fin <- WebCorpus(GoogleFinanceSource("NASDAQ:A
MZN"))
```

The resulting object has three object classes "WebCorpus," "VCorpus" and "Corpus." This means that the object is immediately accepted into tm functions. You can check using the "class" function shown here. Using the "summary" function will print the number of documents contained in the corpus.

```
class(goog.fin)
```

Like all VCorpus class objects, you can access different information with indexing. The first article's content can be accessed in the first line of code and the metadata in the second. To review another article change the double bracketed number.

```
goog.fin[[1]][1]
goog.fin[[1]][2]
```

Another example of the webmining plugin is shown below. This constructs a corpus from a Yahoo news source for the search term "Cleveland." There is no need to specify a specific market as was shown earlier. The different source functions may fail, as the underlying web pages change. If that occurs, use a different method for collecting the text.

```
yahoonews <- WebCorpus(YahooNewsSource("Cleveland"))
```

Once you have your corpus object, you can begin cleaning and analyzing it. If you need to save the documents, you invoke the writeCorpus function, which will save an individual plain text document corresponding to a corpus document. The writeCorpus function is passed the VCorpus object. The function also needs a directory path. Using the period will just write the files to the working directory. Next, the file names have to be specified. Since the goog.fin corpus has 20 documents, the number sequence 1 to 20 is concatenated to goog.fin.txt. In the next section, you will learn how to read multiple files in a directory for text processing. That will allow you to import the 20 individual files quickly to reconstruct the corpus.

```
writeCorpus(goog.fin, path = ".",filenames =
paste0(seq_along(goog.fin), "goog.fin.txt", sep = ""))
```

9.3 Getting Text from File Sources

This section is a good reference as you navigate the numerous file types during a text mining project. Each text mining project could present you with single or multiple file types. This section will serve as an aid so you can easily read files and start the more interesting text analysis. Example files can be downloaded from www.tedkwartler.com download section. The files to be downloaded include

- bos_airbnb_1k.csv
- hillary-clinton-emails-august-31-release.txt
- 1-email.docx
- one_two_star_reviews.xlsx
- turkish_ankara_1.rds
- pdftotext.exe
- hillary-clinton-emails-august-31-release
- IMG_3234.JPG

9.3.1 Individual CSV, TXT and Microsoft Office Files

You have already used the read.csv function but it is presented here for reference. Parsing the common file extension, ".txt," is also presented. The section also includes functions for loading Microsoft Office files such as Excel and Word. The last part of this section demonstrates saving and loading a single RDS file that can be used for any R object not just text.

The read.csv function reads in a comma separated value (CSV) table. The function will create a data frame from the CSV file. The function accepts the file name within quotation marks, along with optional parameters. A common parameter is "header" which can be True or False. Header will determine if the first row contains column names. Another common parameter is "sep," which is for explicitly defining the field separation character. A comma is used most often, but sometimes whitespace or another specific character denotes a new separator. You can also add "stringsAsFactors=" along with a True or False. This determines if character vectors should be strings or categorical variables. The scripts in this book define this option globally for your R session instead of within the file reading step.

```
text<-read.csv('bos_airbnb_1k.csv', headr=T, sep=',')
```

Next to read a .txt file, you can use readLines. This function reads a file line by line to construct a single character vector. You may encounter a warning stating "incomplete final line." Many plain text files end with an end of line (EOL) marker. The warning is letting you know that the function did not encounter one, and the file may be incomplete. This warning can often be ignored after checking the final line. If you want to read a portion of the file, you can add an integer parameter "n=" to read a specific number of lines. The "n" parameter is separated by a comma and includes the number of lines to read. There are other less frequent parameters contained in the function documentation. Since the file was read line by line, the object will contain all the text separated by the EOL markers. If the file is a single document, you may need to use paste and collapse to concatenate the text.

```
emails <- readLines("hillary-clinton-emails-august-
31-release.txt")
emails<-paste(emails, collapse=" ")
```

The qdap library provides one way to gather text from a Microsoft Word document. The `read_docx` function has two inputs. First, the file name to import. The second, "skip=," is an integer parameter similar to the readLines "n" parameter. It defines the number of lines to skip when parsing the Word document. The resulting object is a text vector for the document. The Word document EOL markers define the length of the R character vector, so you can use paste to collapse the text.

```
library(qdap)
one.email <- read_docx('1-email.docx')
one.email<-paste(one.email, collapse=" ")
```

Another common file type is an Excel file. Excel files are called workbooks. Each workbook can contain multiple worksheets with data. Multiple R functions can read in Excel files. The `read.xls` function is from a popular data manipulation library called `gdata`. In addition, functions from the XLConnect library are demonstrated to load an entire workbook or single worksheet.

Using gdata's `read.xls`, you pass in a file name. Then you can pass in the sheet to read. Without a specific sheet, the "read.xls" function will import the first worksheet. This function creates a temporary CSV file for the specific worksheet within the workbook. The temporary file is then read into R. As a result, this function can be slow, especially if you need to load more than one worksheet.

```
library(gdata)
text <- read.xls("one_two_star_reviews.xlsx")
```

The XLConnect library works in a similar manner. Instead of `read.xls` use `readWorksheetFromFile`. In this example, explicitly define the worksheet using "sheet=" and a corresponding integer.

```
library(XLConnect)
text <- readWorksheetFromFile("one_two_star_reviews.
xlsx",sheet = 1)
```

A difference between the gdata and XLConnect package is that the XLConnect library contains a function to load an entire workbook. Then you can create an object for each worksheet within R without repeating code. This can be helpful if you are dealing with a workbook containing many worksheets. However, since this package uses Java it is also memory intensive.

The `loadWorkbook` function is passed an Excel file. At this point, all the data for the entire workbook is held in your R session. The `readWorksheet` function extracts the individual worksheets from the larger workbook.

```
library(XLConnect)
reviews.workbook <- loadWorkbook("one_two_star_
reviews.xlsx")
one.stars <- readWorksheet(reviews.workbook,sheet=1)
two.stars <- readWorksheet(reviews.workbook,sheet=2)
```

The last single file you will learn in this section has an "RDS" extension. This file type is native to R and can be used for any R object. If you are dealing with non-English texts like Arabic tweets, then you should save the object as an RDS file. Saving the object directly from R without changing its character encoding helps ensure that the characters are properly formatted. If you save Arabic tweets as a simple CSV, the characters will be changed. However, saving an RDS file provides direct serialization of the object without altering the characters.

To save any R object, use the saveRDS function, pass in the object and specify a file name ending in ".rds." This will save the object directly to your working directory. Be sure not to overwrite any existing file names in your directory.

```
saveRDS(turk.tweets,'turkish_ankara_1.rds')
```

To load the file into your R console use `readRDS` along with the complete file name. If you need to list all files in your working directory you can call the `dir()` function. This prints all files in the working directory to your console. Then you can copy and paste the exact complete file name into "readRDS."

```
dir()
turk.tweets<-readRDS('turkish_ankara_1.rds')
```

9.3.2 Reading Multiple Files Quickly

So far you have learned how to read common individual files. Yet, you will often have more than one document to load for a text mining project. Reading in many files by copying lines of code and changing files names is tedious. The functions here present two methods for scanning and loading specific files in an efficient manner.

Both approaches use `list.files` to create a vector of file names. The `list.files` function scans your working directory to identify any files that match the specified pattern. The function is similar to using grep for keyword scanning. For example, you create the temp object by searching for any file that ends with "*.csv." The asterisk represents a wildcard meaning a file can have any name, but must end with ".csv." If you are not working with CSV files, you should adjust the pattern to match a different file extension like "*.docx."

```
temp <- list.files(pattern="*.csv")
```

The temp object is a character vector. For example, the temp object may recognize three files matching the pattern such as "allstate_glassdoor.csv", "amzn_cs.csv" and "oct_delta.csv".

Use the following code to load each file as an individual object. Each object is assigned a name corresponding to the file name from your directory. Keep in mind, it will be difficult to call the objects in your console if the object name contains spaces. Thus, it is good practice to use dashes instead of spaces when naming files. The code below is a "for loop" using read.csv. If your files have a different file extension change read.csv to an appropriate importing function.

```
for (i in 1:length(temp)) assign(temp[i], read.
csv(temp[i]))
```

The "for loop" may look confusing because of the "i" variable. The "i" variable changes as the function loops through the temp character vector. So for numbers 1 to the total number of files identified in the temp object, assign the same object name while reading in the CSV. In this example, the function will perform three loops, individually inserting the three file names from the temp object. Your console will now have three distinct data frames called "allstate_glassdoor.csv", "amzn_cs.csv" and "oct_delta.csv".

Instead of an individual object per file the next example binds all files into a single data frame. The code uses the rbindlist function from data.table and a progress bar version of lapply. The code still references the temp object from above. Specifically, the code applies fread over the temp character vector. The fread function is a "fast file finagler" function that you have encountered earlier in the book to parse data tables. The resulting list of data tables is then row bound using rbindlist. Using the rbindlist function along with fill=T, will match columns by name and, if data tables differ, will fill in "NA" for non-matching columns. In this example, you will have a single data frame containing all three CSV files, "allstate_glassdoor.csv", "amzn_cs.csv" and "oct_delta.csv".

```
library(data.table)
library(pbapply)
temp <- list.files(pattern="*.csv")
single.df<- rbindlist(pblapply(temp, fread),fill=T)
```

This solution works with CSV and .txt files that are organized as data tables. You will not be able to change fread to read_docx (or similar) if you are working with other file types. Instead, the code below can be adjusted for other read functions to create a single object containing all documents. Use list.files again and then "pblapply" the appropriate read function to the temp vector. The resulting list can be unlisted to collapse the lines of data into a single object. For example, if "temp" contains "1-email.docx" with 10 lines and "2-email.docx" with 12 lines, the "single.doc.df" object will be a character vector with 22 lines.

```
temp <- list.files(pattern="*.docx")
doc.list<-pblapply(temp, read_docx)
single.doc.df<- unlist(doc.list)
```

9.3.3 Extracting Text from PDFs

The portable document file (PDF) file extension is a widely used Adobe product. PDF files have embedded security and can be hard to edit. As a result, PDF files are great for sending electronically. However, the file security means that extracting the text to analyze can be challenging in R. In fact, you have to load an additional executable program because R cannot do it natively.

To extract text from PDF files, navigate to www.tedkwartler.com, click the "downloads" link and save the file named "pdftotext.exe." This is a copy of the original open source executable file from http://www.foolabs.com/xpdf/download. html. For this example, you can download the PDF named "hillary-clinton-emails-august-31-release.pdf." This is a large PDF containing approximately 7000 emails from Hillary Clinton's time as Secretary of State in the US. Save pdftotext.exe in an easy-to-find directory because you will need to note its file path.

Once you have the executable and one or more pdf files you are ready to extract text. Again use list.files to capture the PDF files, but add an additional parameter. The parameter full.names captures both the file path and name. Next, for the pdf2text object, adjust the file path for your pdftotext. exe download location. The last line of code executes a system level command. Specifically, the small text extracting executable program is passed any files that are recognized from list.files. After using this code, the working directory will contain the original PDF along with a single .txt file.

```
pdf.file <- list.files(pattern = "*.pdf",  full.names
= TRUE)
pdf2text <- "C:/Users/TextMiner/Desktop/pdftotext.exe"
system(paste("\"", pdf2text, "\" \"", pdf.file, "\"",
sep = ""), wait = F)
```

The previous code works for a single file, but needs to be adjusted to work on multiple PDF files in the directory. If your list.files function identifies more than one PDF, an apply function is used to iterate over the multiple file paths. Using pblapply, multiple system level commands are generated for each of the PDF files. Each command separately invokes the pdftotext.exe application with a specific PDF file. The individual commands have the same structure as the single file example. Now your working directory will contain multiple PDF and plain text files.

```
library(pbapply)
pdf.files <- list.files(pattern = "*.pdf",  full.names
= TRUE)
pblapply(pdf.files, function(i) system(paste('"C:/
Users/Edward/Desktop/pdftotext.exe"', paste0('"', i,
'"')), wait = FALSE))
```

9.3.4 Optical Character Recognition: Extracting Text from Images

In rare instances, you may need to extract text from images. Manually transcribing an image is the most accurate method for identifying text in an image. If you do not have time or inclination for such as monotonous task, you can use an outside service for optical character recognition (OCR). OCR services have prebuilt algorithms for identifying letters at the pixel level. Often the algorithms are deep neural nets that are hard to create and computationally intensive. Microsoft provides one such service for automatic OCR with a free and paid tier. Within the Microsoft Cognitive Services there is a computer vision API called Project Oxford. Microsoft's Project Oxford OCR service can be called with an API similar to prior examples. Instead of performing a "get" request to a news service you are now using "post" to send an image for processing. The Microsoft OCR service will provide the text back to your R console although the text needs considerable processing.

To perform OCR using the Microsoft service you will need an API key. Navigate to the Microsoft Cognitive Services webpage or search for Project Oxford API. Then create an account and sign up for the Machine Vision API.

There is an R package, Roxford, in development working to integrate R and Project Oxford. Since Roxford is not officially offered, this section's code demonstrates an API file post using the "httr" library. This example mimics the Roxford developmental package found on www.github.com.

Load three packages to start. The httr package will let you post your file to the service. By now you should be familiar with pbapply, which provides progress bar versions of the apply functions. Lastly, load the plyr library. The plyr package will help manipulate the API's response.

```
library(httr)
library(pbapply)
library(plyr)
```

Next, create three objects. The first is a character string for the Machine Vision API. This is the web address for posting the image file. The URL has two parameters. First, the URL has a True for "detectOrientation" because this helps the algorithm accuracy. Second, the last two letters of the URL represent the language. This example uses English so the link contains "en." If you are working with another language, you can consult the API documentation for the correct abbreviation. Next, use "upload_file" with the image name. Lastly, replace the "x' character string with your Microsoft API key.

```
msft.url <- "https://api.projectoxford.ai/vision/v1/
ocr?detectOrientation=true&language=en"
text.img <- upload_file("IMG_3234.JPG")
api.key <- 'long_string_from_MSFT'
```

Figure 9.9 is the example image to be posted. The image is a portion of a Washington Post article from January 1, 2000. The image contains mostly text along with a portion of a fireworks picture. The API will identify the text and ignore the non-text elements. You can download the file from www.tedkwartler. com or use your own.

The important step in the OCR process is the API POST. The POST command uses the "msft.url", "api.key" and "text.img" objects. Other inputs include the "octet-stream" parameter and a specific header name. Octet-stream dictates that the image file will be sent as a binary file attachment with the other POST information. The header "Ocp-Apim-Subscription-Key" value is specific to Microsoft services and is the name given to your API key credentials.

```
oxford.response <- POST(url = msft.url,
  content_type('application/octet-stream'), add_
headers(.headers = c('Ocp-Apim-Subscription-Key' =
api.key)),body = text.img, encode = 'multipart')
```

Executing this code will send the file, and your API key to the OCR service. To check the results of the API call, type the response object, "oxford.response," into your console. If the file POST was successful, the status will be 200. If there was an issue, the status would be 400 meaning "bad request." The code below shows the console output of a successful API call.

```
oxford.response
#Response [https://api.projectoxford.ai/vision/v1/
ocr?detectOrientation=true]
#  Date: 2016-07-24 03:07
#  Status: 200
#  Content-Type: application/json; charset=utf-8
#  Size: 7.38 kB
```

Httr provides a function, "content," to extract the API response. The api.content is a complicated nested list. Only the subset of the "regions" element contains the text that was recognized and the pixel level box where the text was identified. However, the regions object is another complicated list. So the last line of code applies as.data.frame to the regions' individual "lines" element. The multiple data frames are then row bound using plyr's rbind.fill. This final text.df object contains the recognized text. Unfortunately, the data frame has coordinates for the located text and many "NA" values. As a result, the single data frame needs to be manipulated further.

```
api.content <- content(oxford.response)
regions <- api.content$regions
text.df<- do.call("rbind.fill",
pblapply(regions[[1]]$lines, as.data.frame))
```

The last code section drops the bounding box coordinates and NAs and concatenates the image text. First the columns containing "Box" are dropped from the data frame. Within the index brackets, the grep command is searching for "Box" among the column names. The columns are dropped because the grep function is preceded with a minus. This removes the coordinate information. However, the data frame contains factor levels instead of character strings. Thus, the second line of code changes the data frame text to character strings using pbsapply and as.character. Third, all NA values need to be removed. A simple method for NA removal is to use is.na to identify the NA values along with an empty character ("") as shown. For the API text to match the image, the individual words need to be concatenated by row and then combined into a single plain text document. To perform the row-wise character concatenation use pbapply with 1. The 1 tells the function to apply the "paste" row wise. Change the 1 to 2 if you need to apply a function column wise. The last line of code puts all text rows into a single document. Figure 9.10 is a screenshot of the final cleaned text for comparison to the article image in Figure 9.9.

```
final.text<-text.df[, -grep("*Box*", colnames(text.
df))]
final.text <- pbsapply(final.text, as.character)
final.text[is.na(final.text)] <- ""
final.text<-pbapply(final.text,1,paste,collapse=" ")
final.text<-paste(final.text, collapse=" ")
```

You may notice some extra whitespace and inconsistent spelling. For example, the city "Washing- ton" incorrectly contains a dash. OCR algorithms perform character level recognition. The article used a dash within the characters of "Washington" due to space constraints of the newspaper. The OCR service identified these characters, but did not predict the word in context. Thus, further preprocessing steps may need to be applied to OCR text before beginning your analysis.

Figure 9.9 A portion of a newspaper image to be sent to the OCR service.

```
> final.text
[1] "By DAVID VON DREHLE     Washington Post Staff Writer     Fears of technological disaster
and apocalyptic violence gave way    to giddy global celebration yester-    day as the first day
of the long- anticipated new year swept peace,    fully westward around the planet    and across
the united states.    The year 2000--for generations    a symbol of the distant, dazzling or  d
aunting future-was greeted       with joy and relief by crowds from  Auckland to London to Washing
-    ton and New york. As many as  300,000 people thronged the Mall    despite tight security mea
sures,    Times Square in Manhattan.    Spectacular fireworks illuminat-    ed some of the wo
rld's most fa- mous sites-the pyramids, the Par-    thenon, the Eiffel Tower, Big Ben    and the
washington Monument.    Never was so much exploded for    friendly reasons.                "
```

Figure 9.10 The final OCR text containing the image text.

9.4 Summary

In this chapter you learned methods to extract text from web pages, APIs, individual files and folders containing multiple files. This chapter is meant to illustrate some methods for various text sources, but is by no means exhaustive. The chapter should serve as an aid when confronted with difficult text sources.

Although it is not a case study, you can rely on the code in this chapter to serve you during a text mining project. Within the text mining workflow an important step is identifying the appropriate text. The code in this chapter will let you analyze the plain text from many different sources and file types. In this chapter you learned

- adjustable components of a URL
- automatically constructing URLs for web scraping
- to web scrape a single page
- to web scrape an entire forum
- using an API within an R package to retrieve text
- to set up a connection with Twitter's API to retrieve tweets
- using R without a dedicated package to parse an API's JSON response
- to use the tm package plugin to retrieve a WebCorpus
- to use XML to parse an RSS feed
- simple methods for reading files types csv, .txt, MS Word and Excel documents
- to make a plain text document (.txt) from a PDF
- to use a Microsoft API to identify text in an image file
- two ways to read in multiple files in a folder.

Index

Text Mining in Practice with R, First Edition. Ted Kwartler.
© 2017 John Wiley & Sons Ltd. Published 2017 by John Wiley & Sons Ltd.

Printed and bound by CPI Group (UK) Ltd, Croydon, CR0 4YY

03/10/2023

08124659-0001